# The Ecological Detective

# MONOGRAPHS IN POPULATION BIOLOGY

## EDITED BY SIMON A. LEVIN AND HENRY S. HORN

*Titles available in the series (by monograph number)*

# The Ecological Detective

## Confronting Models with Data

RAY HILBORN

AND

MARC MANGEL

PRINCETON UNIVERSITY PRESS

PRINCETON, NEW JERSEY

1997

Copyright © 1997 by Princeton University Press
Published by PrincetonUniversity Press, 41 William Street,
Princeton, New Jersey 08540
In the United Kingdom: Princeton University Press,
Chichester, West Sussex

**Library of Congress Cataloging-in-Publication Data**

Hilborn, Ray, 1947–
  The ecological detective : confronting models with data / Ray Hilborn
and Marc Mangel.
     p.    cm. — (Monographs in population biology ; 28)
  Includes bibliographical references and index.
  ISBN 0-691-03496-6 (cloth : alk. paper). — ISBN 0-691-03497-4 (pbk. :
alk. paper).
  1. Ecology—Mathematics models. I. Mangel, Marc.   II. Title.
III. Series.
  QH541.15.M3H54   1997     96-9638
  574.5'0151—dc20     CIP

This book has been composed in Baskerville

Princeton University Press books are printed on acid-free paper and
meet the guidelines for permanence and durability of the Committee
on Production Guidelines for Book Longevity of the Council on
Library Resources

Printed in the United States of America

10 9 8 7 6 5 4 3 2 1

10 9 8 7 6
(Pbk.)

http://pup.princeton.edu

ISBN-13: 978-0-691-03497-3 (pbk.)

ISBN-10: 0-691-03497-4 (pbk.)

To Ulrike and Susan
No thanks would be too much

# Contents

CONTENTS

CONTENTS

# CONTENTS

# Preface

## *Beyond the Null Hypothesis*

### ABOUT THE TITLE

First, a word about the phrase "ecological detective," which we owe to our colleague Jon Schnute:

> I once found myself seated on an airplane next to a charming woman whose interests revolved primarily around the activities of her very energetic family. At one point in the conversation came the inevitable question: "What sort of work do you do?" I confess that I rather hate that question. . . . I replied to the woman: "Well, I work with fish populations. The trouble with fish is that you never get to see the whole population. They're not like trees, whose numbers could perhaps be estimated by flying over the forest. Mostly, you see fish only when they're caught. . . . So, you see, if you study fish populations, you tend to get little pieces of information here and there. These bits of information are like the tip of the iceberg; they're part of a much larger story. My job is to try to put the story together. I'm a detective, really, who assembles clues into a coherent picture." (Schnute 1987, 210)

As we began outlining the present volume, we realized that the phrase the "ecological detective" was most appropriate for what we are trying to accomplish. Some reviewers agreed, and some found it a bit too cute. After serious consideration, we decided to leave references to the ecological detective in the text, with apologies to readers who are offended. We find it preferable to "the reader."

It is our view that the ecological detective goes beyond the null hypothesis. As the revolution in physics in the twentieth century showed, there are few cases in science in which absolute truth exists. Models are metaphorical (albeit sometimes accurate) descriptions of nature, and there can never be a "correct" model. There may be a "best" model, which is more consistent with the data than any of its competitors, or several models may be contenders because each is consistent in some way with the data and none clearly dominates the others. It is the job of the ecological detective to determine the support that the data offer for each competing model or hypothesis. The techniques that we introduce, particularly maximum likelihood methods and Bayesian analysis, are the beginning of a new kind of toolkit for doing the job of ecological detection.

## THE AUDIENCE AND ASSUMED BACKGROUND

In a very real way, this book began in October 1988, when we participated in an autumn workshop on mathematical ecology at the International Center for Theoretical Physics. Most of the participants were scientists who had been students in the two previous autumn courses. As these former students presented their work, we realized that although they had received excellent training in ecological modeling and the analysis of ecological models (cf. Levin et al. 1989), they were almost completely inexperienced in the process of connecting data to those models. For scientists in third-world countries, who will work on practical and important problems faced by their nations, such connections are essential, because real answers are needed. We decided then to try to provide the connection.

We envision that readers of this book will be third-year students in biology and upward. Thus, we expect the reader to have had a year of calculus, some classical statistics (typically regression, standard sampling theory, hypothesis testing, and analysis of variance) and some of the classical ecological models (logistic equation, competition equations) equivalent to the material in Krebs's (1994) textbook. Therefore, we will not explain either these classical statistical methods or the classical ecological models. Some readers of drafts took us to task, writing comments such as, "I took my last mathematics and statistics courses four years ago—how dare you expect me to remember any of it." Well, we expect you to remember it and use it. You should not expect to make progress with an attitude of "I learned it once, promptly forgot it, and don't want to learn it again." We worked hard to make the material accessible and understandable, but the motivation rests with you. The more effectively you can deal with data, the greater your contribution to ecology.

This book has equations in it. The equations correspond to real biological situations. There are three levels at which one can understand the equations. The first (lowest) level occurs when you read our explanation of the meaning of the equations. We have tried to do this as effectively as possible, but success can only really be guaranteed in that regard when there is interpersonal contact between student and teacher. The second (middle) occurs when you are able to convert the equation to words—and we encourage you to do so with *every* equation that you encounter. The third (highest) occurs when you explain the origin and meaning of the equation to a colleague. We also encourage you to strive for that.

## COMPUTER PROGRAMMING

Computing is essential for ecological detection. We expect that you have access to a computer. Early drafts of the book, read by many reviewers, had computer programs (rather than pseudocodes) embedded in the text. Virtually all reviewers told us that this was a terrible idea, so we removed them. To really use the methods that we describe here, you must be computer-literate. It does not have to be a fancy computer language: Mangel does all of his work in TrueBASIC and Hilborn does all of his in QuickBASIC or Excel. We recommend that you become familiar with some computer environment that offers nonlinear function minimization, and that you program the examples as you go through the book. For complete neophytes, Mangel and Clark (1988) wrote an introduction to computer programming, focusing on behavioral ecology. In any case, this material will be learned much more effectively if you actually stop at various points and program the material that we are discussing. To be helpful, we give an algorithmic description, which we call a pseudocode, showing how to compute the required quantities. You cannot use these descriptions directly for computation, but they are guides for programming in whatever language you like. Understanding ecological data requires practice at computation, and if you read this book without trying to do any of the computations, you will get much less out of it.

## REALISM AND PROFESSIONALISM

Each of the case studies we use to illustrate a particular point is a bona fide research study conducted by one of us. Even so, some readers of drafts accused us of the unpleasant and unprofessional, but too common (especially in evolutionary biology), behavior of setting up "strawpersons" just to knock them down or of misrepresenting opponent positions (see Selzer 1993 for an example). For example, we were told to

> treat each case study like a real research study and do not spend time rejecting obviously silly models. For example, no one should seriously try to fit a simple logistic equation to the data shown in Figure 8.1. Similarly, one would not need any formal analysis to reject the constant clutch model when presented with the data in Table 6.1. . . .

This would get away from what our class came to call your "toy example" approach—illustrating models or techniques with silly examples, and then not explaining the hard decisions associated with the more interesting and complicated questions.

This charge is unfair. These apparently ridiculous models were *in fact proposed and used by pretty smart people.* Why? Because they had no alternative model. Our view is that the confrontation between more than one model arbitrated by the data underlies science. If there is only one model, it will be used, whether the questions concern management (as in the Serengeti example) or basic science (as in the insect oviposition example). Without multiple models, there is no alternative. Furthermore, in the case studies used here, the data are moderately simple and mainly one-dimensional. This allows us to "eyeball" the data and draw conclusions such as those given above. But in more complicated situations, this may not be possible.

Another side of professionalism is the development of a professional library. As described above, we consider this book a link between standard ecological modeling or theoretical ecology and serious statistical texts. After reading *The Ecological Detective*, the latter should be accessible to you. We consider that a good detective's library includes the following:

Efron, B., and R. Tibshirani. 1993. *An Introduction to the Bootstrap.* Chapman and Hall, New York.

Gelman, A., J. B. Carlin, H. S. Stern, and D. B. Rubin. 1995. *Bayesian Data Analysis.* Chapman and Hall, New York.

McCullagh, P., and J. A. Nelder. 1989. *Generalized Linear Models.* Chapman and Hall, New York.

Press, W. H., B. P. Flannery, S. A. Teukolsky, and W. T. Vetterling. 1986. *Numerical Recipes.* Cambridge University Press, Cambridge, U.K.

PREFACE

ACKNOWLEDGMENTS

We owe lots of people thanks for careful reading, criticisms, and picking nits about the book. For comments on the entire volume, we thank Dick Gomulkiewicz, Peter Turchin, and

- Peter Kareiva and Zoology 572B, spring 1993, University of Washington: Three anonymous students, Beth Babcock, J. Banks, Tamre Cardoso, Kevin Craig, Bill Fagan, Ellen Gryj, Michael Guttormsen, Murdoch McAllister, Jennifer Ruesink (who provided a handy sketch of the ecological detective), Alice Shelly, Dave Somers, Ellen Smith, E. A. Steel (who provided some literature citations we did not know about), Angela B. Stringer, and Colin Wilson.
- Bill Morris and Biology/Zoology 295, fall 1993, Duke University: Sharon Billings, Kerry Bright, Sonia Cavigelli, Chiru Chang, Chris Damiani, Marc Dantzker, Brian Inouye, Brad Klepetka, Sarah May, Yavor Parashkevov, Nathan Phillips, Colleen Rasmussen, George Wilhere, and Neville Yoon.
- Dan Doak and Environmental Studies 270, winter 1995, University of California, Santa Cruz: Elizabeth Andrews, Dave Bigger, Geoff Bryden, Chris Cogan, Bruce Goldstein, Elaine Harding-Smith, David Hyrenbach, Suzy Kohin, Chris Lay, Michelle Marvier, Dan Monson, Jim Murphy, Michelle Paddack, Jacob Pollak, and Laura Trujillo.
- John Gillespie and Population Biology 298, winter 1995, University of California, Davis: Dave Cutler, Arleen Feng, Shea Gardner, Martha Hoopes, Barney Luttbeg, Steve Martin, Lance Morgan, Laura Peck, Rivera Perales, Rich van Buskirk, and Mike Zwick.
- The students in Ray Hilborn's graduate seminar, winter 1995: Eric Anderson, Patricia Dell'Arciprete, Nicholas

xvi

Beer, Claribel Coronado, Bill Driskell, Marianne Johnson, Alan Lowther, Mike McCann, Warren Schlechte, and Adrian Spidle.

We thank Jay Rosenheim and Tim Collier for comments on Chapters 5 and 6.

# The Ecological Detective

# An Ecological Scenario and the Tools of the Ecological Detective

## AN ECOLOGICAL SCENARIO

The Mediterranean fruit fly (medfly), *Ceratitis capitata* (Wiedemann), is one of the most destructive agricultural pests in the world, causing millions of dollars of damage each year. In California, climatic and host conditions are right for establishment of the medfly; this causes considerable concern. In Southern California, populations of medfly have shown sporadic outbreaks (evidenced by trap catch) over the last two decades (Figure 1.1). Until 1991, the accepted view was that each outbreak of the medfly corresponded to a "new" invasion, started by somebody accidentally bringing flies into the state (presumably with rotten fruit). In 1991, our colleague James Carey challenged this view (Carey 1991) and proposed two possible models concerning medfly outbreaks (Figure 1.2). The first model, $M_1$, corresponds to the accepted view: each outbreak of medfly is caused by a new colonization event. After successful colonization, the population grows until it exceeds the detection level and an "invasion" is recorded and eradicated. The second model, $M_2$, is based on the assumption that the medfly has established itself in California at one or more suitable sites, but that, in general, conditions cause the population to remain below the level for detection. On occasion, however, conditions change and the population begins to grow in time and spread over space until detection occurs. Carey argued that the temporal and spatial distributions of trap catch indicate that the medfly may be permanently estab-

3

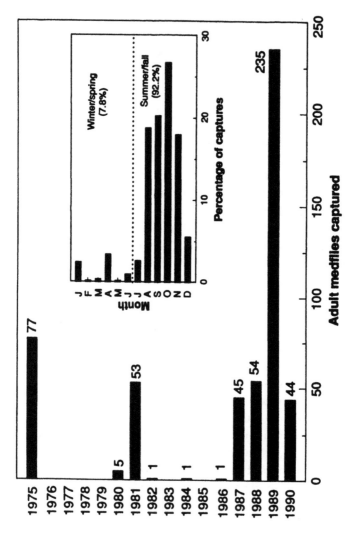

FIGURE 1.1. The capture of adult medfly in Southern California, from 1975 to 1990. (Reprinted with permission from Carey 1991. Copyright American Association for the Advancement of

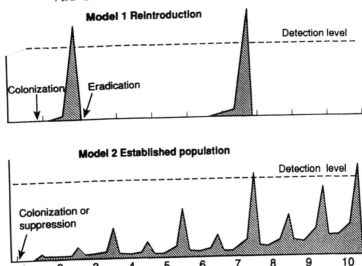

FIGURE 1.2. The outbreak of medfly can be described by two different methods. In model 1, we assume that there is continual reintroduction of the pest. After a reintroduction, the population grows until it exceeds the detection level. In model 2, we assume that the medfly is established, but that ecological conditions are only occasionally suitable for it to grow and exceed the detection threshold. (Reprinted with permission from Carey 1991. Copyright American Association for the Advancement of Science.)

lished in the Los Angeles area. Knowing which of these views is more correct is important from a number of perspectives, including the basic biology of invasions and the implications of an established pest on agricultural practices.

Determining which model is more consistent with the data is a problem in ecological detection. That is, if we allow that either model $M_1$ or model $M_2$ is true, we would like to associate probabilities, given the data, with the two models. We shall refer to this as "the probability of the model" or the "degree of belief in the model." How might such a problem be solved? First, we must characterize the available data, which are the spatial distribution of trap catches of medfly over time (Figure 1.3). We could refine these by placing

5

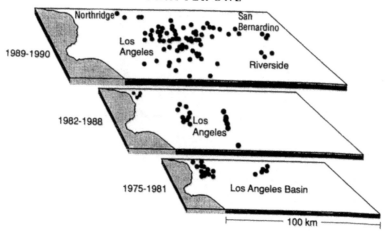

Figure 1.3. The data available for ecological detection in this case would be the spatial distribution of the catch of adult medfly over time. (Reprinted with permission from Carey 1991. Copyright American Association for the Advancement of Science.)

small grids over the maps and characterizing a variable that measures the number of flies that appear in cell $i$ in year $y$. Second, we must convert the pictorial or verbal models shown in Figure 1.2 into mathematical descriptions. That is, some kind of mathematical model is needed so that the data can be compared with predictions of a model. Such models would be used to predict the temporal and spatial patterns in detected outbreaks; the mathematical descriptions would generate maps similar to the figures. The models would involve at least two submodels, one for the population dynamics and one for the detection process. Courses in ecological modeling show how this is done. Third, we confront the models with the data by comparing the predicted and observed results. At least three approaches can be broadly identified for such a confrontation.

*Classical Hypothesis Testing.* Here we confront each model separately with the data. Thus, we begin with hypotheses:

6

is true

model is true

ight be that outbreaks are
sing the mathematical de-
struct a "$p$ value" for the
happen that we can defi-
e is so small (usually less
, we might not be able to
v.05), but then might discover that the
..... of the statistical test is quite low (we assume that most
readers are probably familiar with the terms "$p$ values" and
"power" from courses in elementary statistics, but we shall
explain them in more detail in the following chapters). In
any case, we use such hypothesis testing because it gives the
"illusion of objectivity" (Berger and Berry 1988; Shaver
1993; Cohen 1994).

After we had tested the hypothesis that model 1 is true
against the alternate hypothesis, we would test the hypothesis that model 2 is true against the alternate. Some of the
outcomes of this procedure could be: (i) both models $M_1$
and $M_2$ are rejected; (ii) model $M_1$ is rejected but $M_2$ is not;
(iii) model $M_1$ is not rejected but $M_2$ is; and (iv) neither
model is rejected. If outcome (ii) or (iii) occurs, then we
will presumably act as if model $M_1$ or $M_2$ were true and
make scientific and policy decisions on that basis, but if outcome (i) or (iv) occurs, what are we to do? Other than col-
lecting more data, we are provided with little guidance concerning how we should now view the models and what they
tell us about the world. There is also a chance that if outcome (ii) or (iii) occurs, the result is wrong, and then on
what basis do we choose the $p$ level?

*Likelihood Approach* (Edwards 1992). In this case, we use
the data to arbitrate between the two models. That is, given
the data and a mathematical description of the two models,
we can ask, "How likely are the data, given the model?" Details of how to do this are given in the rest of this book, but

7

read on pretending that you indeed know
Thus, we first construct a measure of the prob
observed data, given that the model is true—we
note this by Pr{data|$M_i$}. We then turn this on its h
interpret it as a measure of the chance that the model
appropriate description of the world, given the data. T
is called the likelihood and we denote it by $\mathscr{L}_i\{M_i|data\}$.
We now compare the likelihoods of the two models, given
the data. If $\mathscr{L}_1\{M_1|data\} \gg \mathscr{L}_2\{M_2|data\}$, then we would
argue that model $M_1$ is a better description of the world;
if $\mathscr{L}_1\{M_1|data\} \ll \mathscr{L}_2\{M_2|data\}$, then we would argue
that model $M_2$ is a better description of the world; and if
$\mathscr{L}_1\{M_1|data\} \sim \mathscr{L}_2\{M_2|data\}$, then we would argue that the
data do not differentiate between the models. A smart deci-
sion maker would not act as if the most likely model were
true, but would weigh the costs and consequences of each
action against the relative probabilities of the alternative hypotheses. But what exactly is meant by "$\gg$," "$\ll$," or "$\sim$" in
this approach?

In this book, we shall work out methods for determining
when one likelihood is much larger than another, and what
that means in terms of confronting models with data.

*Bayesian Approach.* Finally, we might have other information that allows us to judge a priori which model is more
likely to be true. For example, we might know the ecology of
invasion and establishment of medfly in other places. Or we
might know that before certain outbreaks people had been
caught bringing fruit into the country from places where
medfly is established. This kind of information can be summarized in a "prior probability that model $M_i$ is true," which
we denote by $p_i$. If we allow only two models of the world
(medfly are established or they reinvade), then $p_1 + p_2 =$
1. Now, given information consisting of trap catches and the
mathematical model, we want to "update" these prior probabilities. That is, we want to evaluate a "posterior probability
that model $M_i$ is true, given the data." Procedures for doing

this require an understanding of conditional probability and are generally called "Bayesian methods," named after the Reverend Thomas Bayes, who introduced such ideas. In biology and mathematics, one of the earliest modern proponents was Sir Harold Jeffreys (1948), who called the method "inverse probability." His goal was to find methods that allow us to combine prior information with the chance of observing the data to evaluate a posterior probability of different hypotheses, given a scenario associated with the prior information. Interestingly, although Jeffreys is most famous for his work in applied mathematics, astronomy, and geophysics, he was one of the earliest contributors to the *Journal of Ecology* (Sheail 1989). In this book, we shall illustrate how Bayesian methods can be developed and applied. They are particularly appropriate for cases in which studies cannot be replicated (e.g., Reckhow 1990) and for assessment of the risk and safety in various environmental settings in which "expert opinion" is sought (Emlen 1989; Apostolakis 1990; Bolt 1991). There are arguments that Bayesian reasoning is the only way to provide a unified and consistent approach to deterministic and statistical theories (Howson and Urbach 1989, 1991).

This ecological scenario illustrates three approaches that can be taken when confronting models with data. It is our opinion that the process of science consists of confronting more than one description of how the world works with data. Thus, in the rest of the book we spend little time on classical methods of hypothesis testing but focus on likelihood and Bayesian methods. Two recent special features in the journal *Ecology* contain a number of papers that deal with nonclassical approaches to the use of statistics in ecological problems (Carpenter 1990; Jassby and Powell 1990; Reckhow 1990; Walters and Holling 1990; Potvin and Roff 1993; Potvin and Travis 1993; Shaw and Mitchell-Olds 1993; Trexler and Travis 1993) or with particularities of ecological situations (Dutilleul 1993; Legendre 1993). They provide a good complementary background for this book.

## THE TOOLS FOR ECOLOGICAL DETECTION

The modern ecologist usually works in both the field and laboratory, uses statistics and computers, and often works with ecological concepts that are model based, if not model driven. How do we make the field and laboratory coherent? How do we link models and data? How do we use statistics to help experimentation? How do we integrate modeling and statistics? How do we confront multiple hypotheses with data and assign degrees of belief to different hypotheses? How do we deal with time series (in which data are linked from one measurement to the next) or put multiple sources of data into one inferential framework? These are the kinds of questions asked and answered by the ecological detective.

Like all other forms of creative activity, ecological detection is a craft that requires the right tools as well as the skills and materials to use the tools. We envision four components.

*Hypotheses* are the first component. Notice the plural, which is essential to our viewpoint. Science consists of confronting different descriptions of how the world works with data, using the data to arbitrate between the different descriptions, and using the "best" description to make additional predictions or decisions. These descriptions of how the world might work are hypotheses, and often they can be translated into quantitative predictions via models. In Chapter 2, we review different kinds of models, the purposes of models, and how models are related to hypotheses.

*Data* are the second component. You cannot do good analysis if the data are not good. But what does "good" mean? Sometimes the role of analysis is to show that a set of data—at least within the context of a particular view of the world—is not as informative or as useful as one thought it would be. In Chapter 3, we stress that it is important to "Know Your Data" and we provide a sufficient review of probability and the stochastic processes that you will need to conduct the work of the ecological detective.

*Goodness of fit* is the third component. When the data are used to arbitrate between different hypotheses or models, we must have a measure to determine how well each description of the world fits the observations. In Chapters 5, 7, and 9, we describe a variety of measures of goodness of fit that can be used in the confrontation of models and data. We provide recommendations about when it is good to use a particular method.

*Numerical procedures* are the fourth component. Having a measure of goodness of fit between the model and the data is not enough—you must to be able to evaluate it quickly and efficiently and explore the goodness of fit of other models. Thus, in Chapter 11, we provide an introduction to numerical methods needed to assess goodness of fit and to find the best fit. There is a history of the use of numerical procedures in ecology (examples from a generation ago are given by Conway et al. 1970, Melzer 1970, and Marten et al. 1975), but it is the development of microcomputers that really allows the full richness of numerical procedures to be exploited by practicing ecologists.

Overarching these components are alternative views of the scientific method and the role of models in science, which we discuss in Chapter 2. There we present four of the major philosophies of science and show how two of them are closely connected to our work of ecological detection.

A final warning. We are practicing ecologists. We are not statisticians, numerical analysts, or philosophers, and the appropriate chapters will no doubt offend the appropriate experts. For this we make no apologies other than stressing that for the ecological detective the problem is paramount. Because of that, we bring to the problem whatever techniques—from wherever they come—needed to solve it. And if the techniques do not exist, then we must invent them.

# Alternative Views of the Scientific Method and of Modeling

Science is a process for learning about nature in which competing ideas about how the world works are measured against observations (Feynman 1965, 1985). Because our descriptions of the world are almost always incomplete and our measurements involve uncertainty and inaccuracy, we require methods for assessing the concordance of the competing ideas and the observations. These methods generally constitute the field of statistics (Stigler 1986). Our purpose in writing this book is to provide ecologists with additional tools to make this process more efficient. Most of the material provided in subsequent chapters deals with formal tools for evaluating the confrontation between ideas and data, but before we delve into the methods we step back and consider the scientific process itself. No scientist can be truly "neutral." We all operate within a fundamental philosophical worldview, and the types of statistical tools we employ and the types of experiments we do depend on that philosophy. Here we present four such philosophies.

There is a commonly accepted model for the scientific process (and from it arose a well-developed body of statistics that is taught in nearly every university in North America). The basic view can be thought of as a learning tree of critical experiments, which was described by Platt (1964) as "strong inference," and consists of the following steps:

1. Devising alternative hypotheses
2. Devising a crucial experiment (or several of them) with

alternative possible outcomes, each of which will, as nearly as possible, exclude one or more of the hypotheses

3. Carrying out the experiment so as to get a clean result
4. Recycling the procedure, making subhypotheses or sequential hypotheses to refine the possibilities that remain, and so on (Platt 1964, 347)

Platt likens this to climbing a tree, where each fork of the tree corresponds to an experimental outcome, and we base the direction of the climb on the outcomes so far. It is especially interesting for us as ecologists that Platt associates a "second great intellectual revolution" with the "method of multiple hypotheses," and attributes some of the most original thinking in this area to the geologist T. C. Chamberlain who published at the end of the last century. In particular, Chamberlain stressed that we are guaranteed to get into trouble when we consider only a single hypothesis rather than multiple hypotheses. This is especially interesting because the similarities between the geological and ecological sciences are in some ways much greater than the similarities between the other physical and the ecological sciences. In both ecology and geology, experiments may be difficult to perform and so we must rely on observation, inference, good thinking, and models to guide our understanding of the world. In fact, ecology may be much more of an "earth science" than a "biological science" (Roughgarden et al. 1994). We include a reprint of Chamberlain's classic paper—first published in the 1890s—as the Appendix.

ALTERNATIVE VIEWS OF THE SCIENTIFIC METHOD

Platt's view is to a very great extent the logical extension of the work of Karl Popper (1979), who revolutionized the philosophy of science in the twentieth century by arguing that hypotheses cannot be proved, but only disproved

13

TABLE 2.1. Four philosophies of science.

| Philosopher | Key word or phrase | Type of confrontation |
|---|---|---|
| Popper | Falsification of hypotheses | Single hypothesis is disproved by confrontation with the data. |
| Kuhn | Paradigms, normal science, scientific revolution | Single hypothesis used until there is so much contradictory information that it is "overthrown" by a "better" hypothesis. |
| Polanyi | Republic of science | Multiple views of the world allowed according to the different opinions of scientists. Confrontation between these views and the data judged on (i) plausibility, (ii) value, (iii) interest. |
| Lakatos | Scientific research program | Confrontation of multiple hypotheses with data as arbitrator. |

(Table 2.1). The essence of Popper's method is to challenge a hypothesis repeatedly with critical experiments. If the hypothesis stands up to repeated experiments, it is not validated, but rather acquires a degree of respect, so that in practice it is treated as if it were true. Most "modern" scientific journals adopt this approach, even though there are difficulties in using it even under the best circumstances (e.g., Lindh 1993).

Coinciding with Popper's philosophical development was the statistical work of Ronald Fisher, Karl Pearson, Jerzy Neyman, and others, who developed much of the modern statistical theory associated with "hypothesis testing" (e.g., Kendall and Stuart 1979, 175 ff.). In hypothesis testing, we focus on a single hypothesis (called the "null hypothesis") and calculate the probability that the data would have been observed if the null hypothesis were true. If this probability is small enough (usually 0.01 or 0.05), then we "reject" the null hypothesis. To complete the calculation, we must also compute the statistical power associated with the test (Peterman 1990a,b; Greenwood 1993; Thompson and Neill 1993). The power is the probability that if the null hypothesis were actually false and we were given the same data, we would reject it.

For example, we might begin with the idea that larger flocks of birds forage more effectively than smaller flocks. The null hypothesis could be that there is no relationship between flock size and foraging efficiency. A typical application of hypothesis testing would be to use linear regression to test the null hypothesis by calculating the probability that the slope of a graph of flock size versus feeding efficiency is non-zero. If the probability that the data could have arisen from the null hypothesis (slope = 0) is greater than 0.05 (or 0.01), the null hypothesis is not rejected at the "5% level" (or the 1% level). In the case considered here, if the null hypothesis could not be rejected at the 5% or 1% level and the power were sufficiently high, then the real ecological hypothesis—larger flocks forage more efficiently—would effectively be rejected.

After testing the hypothesis that larger flocks forage more efficiently, we would continue to climb Platt's decision tree to another set of experiments, depending on whether the effect of flock size on foraging efficiency was or was not statistically significant. The key elements of this view of science are (1) the confrontation between a single hypothesis and the data, (2) the central idea of the critical experiment, and (3) falsification as the only "truth." Popper supplied the philosophy and Fisher, Pearson, and colleagues supplied the statistics. At best, this view of science is exceptionally narrow and actually does not fit many ecological situations. At worst, it can be downright dangerous, if, for instance, we accept the null hypothesis as true and the experiment had low power (also see Bernays and Wege 1987). Before we explain our own perspective, we want to provide an overview of some other views of science.

Thomas Kuhn (1962) introduced the ideas of "normal science," "scientific paradigms," and "scientific revolutions." According to Kuhn, scientists normally operate within specific paradigms, which are broad descriptions of the way nature works. Normal science involves collection of data within the context of the existing paradigm. Normal science does not confront the existing paradigm, rather, it embellishes it. The paradigm dictates what type of experiments to perform, what data to collect, and how to interpret the data. In Kuhn's view, real change occurs only when (i) a large body of contradictory data accumulates and the existing paradigm cannot explain the data, and (ii) there is an alternative paradigm that can explain the discrepancies between the old paradigm and the observations. Kuhn argues that there is rarely, if ever, a critical experiment at the level of the paradigm. Instead, a particular anomaly will be explained as a measurement problem. It is the collection of contradictory experiments that leads to the revolution.

The Kuhnian perspective is that the type of experimental trees and critical experiments described by Platt may occur, but only within an individual paradigm, and that they are the standard procedures of normal science. The example we gave earlier of examining the relationship between flock size and foraging efficiency would be considered normal science within a broad paradigm of natural selection acting on behavior.

Michael Polanyi (1969) describes a "republic of science" consisting of a community of independent thinkers cooperating in a relatively free spirit. To Polanyi, this represents a

16

simplified version of a free society in which scientists develop by "training" with a "master" so that the practice of science is analogous to apprenticing with a master artisan and learning the skills of the artisan by close observation and participation. Scientists are chosen through this apprenticing system; the individuals constitute the "republic" of citizens taught through the master-apprentice chain. It is this system that prevents science from becoming moribund or rigid, since the apprentice both learns high standards from the master and develops his or her own judgment for scientific matters. There are three main criteria for judgment (Polanyi 1969, 53 ff.): (1) plausibility, (2) scientific value (consisting of accuracy, intrinsic interest, and importance), and (3) originality. The criteria of plausibility and scientific value will encourage conformity, whereas the value given to originality encourages creative thinking and dissent. This forms the essential tension in any scientific field, and the three criteria considered by Polanyi are appropriate ones that we can use for confronting models with data. Polanyi implicitly argues that the intellectual confrontation is not between a model and data, but between models (i.e., different descriptions of how the world works) and data (observations and measurements).

There is an overlap between the ideas of Polanyi and Kuhn. The apprentice system is the essence of Kuhn's normal science: apprentices learn from their masters what type of experiments to perform, and then, to a large extent, continue to work on this type of problem for the rest of their careers. It is the unusual scientist who breaks away from the material of the apprenticeship and enters a new field. We have noticed how common it is in ecology for someone to do a Ph.D. in a specific area, often with a particular taxonomic group, and then continue for most of a career to study the same topic. One of our colleagues in a chemistry department said that it was the same in his field: more than 70% of his colleagues worked with the same types of reactions they studied for their Ph.D.'s. This is the apprentice system and normal science. It is unlikely to lead to innovation or breakthroughs.

Imre Lakatos (1978) describes "scientific research programs" (SRPs) that consist of a set of methodological rules that guide research by indicating paths to avoid and paths to pursue. The "hard core" is the key element of the SRP, which generates a set of surrounding hypotheses that make specific predictions. Lakatos refers to these surrounding hypotheses as a "belt" that protects the hard core. The individual elements of the belt can be tested, and rejected, but one can rarely, if ever, directly challenge the hard core.

Lakatos points out that many hypotheses (e.g., Newton's laws and the theory of gravity) have been highly regarded and used despite their acknowledged inconsistency with some aspect of the data. Organic chemists worked for years with models that they knew were wrong but for which alternatives were lacking. Lakatos argues that the value of an SRP is its ability to make new predictions and provide a simple and elegant explanation of what is known. An SRP can only be replaced by another SRP: One cannot reject a hypothesis unless there is something better on hand to replace it. Mitchell and Valone (1990) argued that optimization in biology should be viewed as an SRP (also see Orzack and Sober 1994).

Thus, in the Lakatosian view, the contest must always be between competing hypotheses and the data. An individual hypothesis may well be inconsistent with the data, but unless there is another hypothesis that is more consistent with the data, you will not discard the first hypothesis because you have to keep working. The recognition of the importance of more than one model is slowly appearing. For example, Chen et al. (1992) compare a number of functions used to describe the growth of fish. If we only consider one growth function, we shall surely use it to make predictions, regardless of its efficacy, but comparing different growth functions allows choice in the description of how nature works. Similarly, Schnute and Groot (1992) confront ten different models of animal orientation with data, Ribbens et al. (1994) compare different models for seedling recruitment in forests, and Kramer (1994) compares six different models for the onset of growth in the European beech.

To a great extent, Popper's view of falsification, Kuhn's normal science, Polanyi's republic, and Lakatos's testing of the "belt" of auxiliary hypotheses are different descriptions of the same scientific activity. It is rare that the major ideas, such as evolution by natural selection or the theory of relativity, are truly tested. In fact, most of the work of the ecological detective will be at a considerably more mundane level. Indeed, it is safe to say that we are writing this book as a handbook for the practice of normal science. (Although, of course, we hope that something more exciting comes from it.)

As briefly described in the previous chapter, the field of likelihood/Bayesian statistics is well suited for the analysis of the contest between competing hypotheses and data. The essence of likelihood/Bayesian analysis is the calculation of the chance of the data given a particular hypothesis, and (for Bayesian methods) from that, "posterior distributions" that describe the probability assigned to each possible hypothesis after data are collected. We describe the mechanics of Bayesian statistics in succeeding chapters. Here we briefly contrast the approaches of classical and likelihood/Bayesian statistics. We shall show in succeeding chapters that likelihood methods are a special case of Bayesian ones, so that from now on we simply refer to them as Bayesian methods.

In classical statistics, we test each hypothesis against the data in a mock confrontation with a "null hypothesis." In Bayesian statistics, we test the hypotheses together against each other, using the data to evaluate the degree of belief that should be accorded each of the hypotheses. The result of a classical analysis is rejection or nonrejection of the lone hypothesis, whereas the result of a Bayesian analysis is "degrees of belief" associated with the different hypotheses.

Two of the three pillars of the classical viewpoint, falsification and the confrontation between a single hypothesis and data, are directly opposed by the Bayesian viewpoint. In the classical approach hypotheses are falsified (but never proved), but in the Bayesian viewpoint degrees of belief are increasing or decreasing. "Falsification" exists only as low degrees of belief and "proof" is strong belief. The two views also are diametrically opposed on whether the confrontation is between a hypothesis and the data, or between competing hypotheses and data. According to Lakatos, we cannot reject a hypothesis unless something better awaits, and Bayesian computation requires more than one hypothesis. In the viewpoint of Popper and classical statistics, we can reject a hypothesis by itself in single combat with data. But then what?

There is much more compatibility between the differing viewpoints on the question of critical experiments. To a Bayesian, a critical experiment is one that will greatly change the degrees of belief in competing hypotheses. Indeed, there is no point in conducting an experiment that will not change the degrees of belief. To a Bayesian, the ideal Popperian critical experiment is one that will change the degrees of belief to almost 1.0 for one hypothesis and almost 0.0 for the others, depending upon the outcome of the experiment. The best experiments are those that discriminate most clearly, although the Popperian/classical view would not require that there be competing hypotheses. We find the Lakatosian/Bayesian view more compelling: that the contest is between competing hypotheses and data, not between a single hypothesis and the data.

We must also consider the issue of statistical significance versus biological significance. Too many people operate on the premise that if statistical significance cannot be shown, the work cannot be published. Yet even elementary statistics courses teach us that statistical significance often has little, if any, relation to biological significance. Two curves can be statistically significantly different even if they differ by less than one percent, given a large enough sample size or small enough measurement error. Conversely, given small sample sizes or high variability, even the most different of biological relationships can fail to be statistically significant. And yet, especially when experiments are difficult or management actions needed, we may not have the luxury of obtaining statistical significance before needing to act on our hypotheses.

## STATISTICAL INFERENCE IN EXPERIMENTAL TREES

Now let us return to Platt's experimental tree and consider it from the different perspectives. The basic structure of an experimental tree is compatible with the varying viewpoints if they are suitably modified. Lakatos would insist that each experiment be a contest between competing hypotheses, whereas Popper would accept experiments testing a hypothesis with no competitor. More importantly, Lakatos would not accept that the "hard core" of an SRP could be experimentally tested in this way. Popper would see the experiments as testing the key hypothesis, since a good hypothesis is one that is amenable to direct experimental falsification.

Platt's experimental tree is based on the premises of (i) very clear and distinct hypotheses and (ii) nonambiguous outcomes. Examining the nature of the statistical tests that could be used in working through an experimental tree shows the problems of the method of hypothesis testing. Imagine you are at experiment A and are asking if larger flocks forage more efficiently. Suppose that if the null hypothesis cannot be rejected, experiment B is appropriate, whereas if the null hypothesis is rejected (therefore large flocks do forage more efficiently), experiment C will be next. What significance level should one choose to decide which branch of the tree to follow? Should experiment C be

21

next, even if the estimated increase in foraging efficiency for larger flocks is biologically trivial, although statistically significant?

In our view, an experimenter would more profitably operate as follows. At the conclusion of experiment A there are really seven options, not two:

1. go on to experiment B,
2. go on to experiment C,
3. repeat experiment A,
4. perform both B and C,
5. perform both A and C,
6. perform both A and B, or
7. perform A, B, and C!

Indeed, if the experiments are inexpensive to set up and run but require considerable waiting time for the outcome, it would be best to do A, B, and C simultaneously.

Progress through an experimental tree thus depends on several factors including (1) the cost of each experiment, (2) the time required to do each experiment, and (3) the relative degree of belief in competing hypotheses. At any stage in the tree, a good scientist will compare the cost and time required to do each experiment to the degree of belief in competing hypotheses and from these calculate the optimal next experiment(s).

## UNIQUE ASPECTS OF ECOLOGICAL DATA

Platt envisioned very clean experiments in which one hypothesis would be clearly discredited. Indeed, a key thrust of Platt's argument is that the fields that made the most rapid progress were those fields that routinely thought about and designed such experiments. Clearly, a field will make more rapid progress if such clear, critical experiments can be designed and conducted, and ecologists should seek to work on systems that are amenable to such analysis. Whenever possible, conduct an experiment (Hairston, 1989, 1994; Underwood, 1991). However, many ecological studies are motivated by problems where such clear experimentation and "hard data" are often not possible (Fagerström 1987) or lead to other difficulties, as the recent "Frontiers in Biology" in *Science* (269:313–61, 1995) and associated correspondence (269:561–64, 1201–3) demonstrate.

For example, consider the problems in understanding the dynamics of populations of blue whales. There is no possibility for experimental manipulation (for decades at least), there is no possibility for replication, since there are so few individuals and they may constitute a single population, and the time scale of their dynamics is very slow. We cannot design a Platt-type experimental tree for manipulation of blue whales—but we could design such an experimental tree for many hypotheses and use observation, rather than experiment, to differentiate between the hypotheses.

Blue whales are an extreme example, but the following attributes of ecological systems often make experimentation difficult:

- Long time scales: Many ecological systems have time scales of years or decades
- Poor replication: Many ecological systems are difficult to replicate, and replicates are rarely, if ever, perfect
- Inability to control: One can rarely, if ever, control all aspects of an ecological experiment

Because of these factors it is often harder to get clear, unambiguous results in ecological experiments (cf. Shrader-Frechette and McCoy 1992). Platt described an experimental approach that did not really need statistics, because each experiment produced a clear result. This is not often the case in ecological work.

Of course, new students should seek systems that do not have these problems, and we encourage you (especially graduate students) to find systems that operate on short time scales and can be easily replicated and easily controlled. It sometimes happens that we are able to apply knowledge from small-scale experimental systems to larger-scale "real world" systems, but it is likely that at least some of the work of the ecological detective will be on ecological systems that may present all three of these difficulties.

## DISTINGUISHING BETWEEN MODELS AND HYPOTHESES

We begin by trying to sort out "theory," "hypothesis," and "model." The etymology of theory is Greek, *theoria*, meaning "a looking at, contemplation, speculation," and we understand theory to mean "a systematic statement of principles involved" or "a formulation of apparent relationships or underlying principles of certain observed phenomena which has been verified to some degree." The theory of evolution by natural selection, without doubt the most important theory in modern biology, is still mainly nonmathematical. The same is true of the theory of Crick and Watson that DNA is a double helix (Crick 1988). The etymology of hypothesis is also Greek, *hypotithenai*, meaning "to place under."

A hypothesis is "an unproved theory, proposition, supposition, etc., tentatively accepted to explain certain facts or to provide a basis for further investigation." Webster's dictionary (Neufeldt and Guralnik 1991) separates theory and hypothesis as follows: "**theory**, as compared here, implies considerable evidence in support of a formulated general principle, explaining the operation of certain phenomena; **hypothesis** implies an inadequacy of support of an explanation that is tentatively inferred, often as a basis for further experimentation."

The etymology of model is from Latin *modus*, meaning the way in which things are done. A model is an archetype, "a stylized representation or a generalized description used in analyzing or explaining something." Thus, models are tools for the evaluation of hypotheses (our best understanding of how the world works), but they are not hypotheses (cf. Caswell 1988; Hall 1988; Onstad 1988; Ulanowicz 1988).

Most hypotheses could be represented by a number of models. The hypothesis that birds forage more efficiently in flocks than individually could be represented by several models relating consumption rate $C$ and flock size $S$:

$C = aS$ Model A: Consumption proportional to flock size,

$C = \dfrac{AS}{1 + bS}$ Model B: Consumption saturates as flock size increases,

$C = aSe^{-bS}$ Model C: Consumption increases and then decreases with increasing flock size, (2.1)

where $a$ and $b$ are parameters of the models. Each model is a more explicit statement of the hypothesis that "birds forage more efficiently in larger flocks" (Figure 2.1). The "null hypothesis" is the model that forage efficiency is independent of flock size, or $C = a$. In the Popperian confrontation models A, B, and C would individually be "tested" against the null hypothesis. In a Lakatosian world the confrontation would be between the four competing models (A, B, C, and the "null").

One can think of hypotheses and models in a hierarchic fashion with models simply being a more specific version of a hypothesis. Furthermore, particular parameter values of the models are even more specific hypotheses. Indeed, in later chapters that deal with probability, likelihood, and Bayes' theorem, we use the word "hypothesis" to refer to particular parameter values of specific mathematical models. The use of "hypothesis" with reference to probabilities is unfortunate, though necessitated by the general statistical usage, but do not confuse the distinction between a hypothesis as a general statement about the natural world and the

25

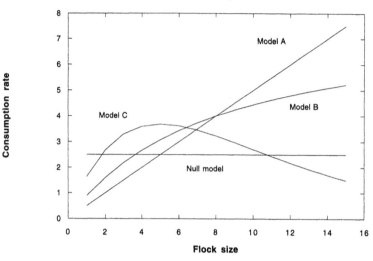

FIGURE 2.1. Three models of how foraging efficiency might be affected by flock size. The flat line is the null hypothesis that flock size does not affect foraging efficiency.

variety of mathematical models that can be used to represent the hypothesis.

We use models to evaluate hypotheses in terms of their ability both to explain existing data and predict other aspects of nature. We use models to combine what we know with our best guesses about what we do not know. The equations of a model represent a very specific expression of the hypothesis. For example, a hypothesis might be that "predation has a significant effect on the average abundance of the population of $X$." Models of this hypothesis would describe the interaction between the organism and its predators in the context of specific mathematical forms (one of which— the null model—could include no predation). Were such models confronted with abundance data, we might find that models including predation explained the abundance of $X$ no better than a model without predation. We would then have some evidence that the hypothesis is incorrect. In this "Lakatosian" view of hypotheses and models, the individual models are the surrounding belt that defends the core hypothesis. We chip away at the individual model and eventually, as we exhaust the possibilities of different mathematical representations of predation, decrease belief in the underlying hypothesis of the importance of predation and increase belief in the alternative hypotheses. Wise (1993) provides an example of how this program is followed in understanding the roles of spiders in ecological systems.

Models have a number of different purposes in the general evaluation of scientific hypotheses. First, models help us clarify verbal descriptions of nature and of mechanisms. Formulation of a model often forces the researcher to think about processes that he or she had previously ignored. The formulation leads to identification of parameters that must be measured and often helps crystallize thinking about the processes involved.

Second, models often help us understand which are the important parameters and processes and which ones are not important. For example, in the formulation of a model we often see that combinations of parameters, rather than the individual parameters themselves, determine the behavior of the system (see Mangel and Clark, 1988, epilogue). Models thus allow us to rank the importance of different factors about the phenomenon in a quantitative manner.

Third, since a model is not a hypothesis we must admit from the outset that there is no "fully correct" model. Instead, there are sequences of models, some of which may be better than others as tools for understanding the natural world. Different models of the same phenomenon can be quite useful, as we shall see in several of the case studies presented later. Different models allow us to assess the validity of different assumptions and, in some cases, of fully different hypotheses. The development of different models usually represents a progression in the understanding of the natural system. This is especially important; one must focus on the system of interest and be willing to forego the model

BOX 2.1

SEPARATING HYPOTHESES AND MODELS: A SCENARIO FROM
CLASSICAL PHYSICS

Here we expand upon an example used by Mangel and Clark (1988). This example requires elementary physics. Envision a mass $M$ attached to a spring which is then attached to a ceiling. We pull the ball away from the ceiling and let it go; the ball starts to oscillate. Our goal is to understand what is happening. We begin with the usual hypothesis of Newton's second law of motion: $F = Ma$ (force equals mass times acceleration). If $X(t)$ denotes the displacement of the mass from the original resting spot at time $t$, then the simplest model for the restoring force is that it is proportional to the displacement

$$\text{Model 1:} \quad M = \frac{d^2X}{dt^2} = -KX. \tag{B2.1}$$

Here $d^2X/dt^2$ is the acceleration. The solution of this differential equation (which you may have once studied in physics or calculus) leads to two important predictions. First, this simple model predicts that the spring will oscillate forever. Second, the frequency of oscillations depends on the combination $\sqrt{K/M}$ and not on $K$ or $M$ independently; this is something that we could not have determined without the model.

However, there are problems. Real springs ultimately slow down and stop oscillating. Do we conclude that the hypothesis $F = Ma$ is wrong or that the model is missing something? For example, we have ignored frictional forces which tend to slow things down according to the size of their velocity. Hence, we might modify model 1 to obtain

$$\text{Model 2:} \quad M\frac{d^2X}{dt^2} = -KX - K_1V, \tag{B2.2}$$

where $V = dX/dt$ is the velocity and we have added another parameter $K_1$ that relates the frictional force and velocity.

BOX 2.1 CONT.

Once again, by solving the equation we could learn that the answer does not depend on $K_1$ itself but on the ratio $K_1/M$, and that model 2 predicts that the spring will slow down. Consequently, this is a clear improvement in the model without any change of hypothesis.

However, real springs slow down and stop in a finite time, but the spring described by model 2 will only stop as time becomes infinite. Once more, we conclude that there is a problem with the model and might introduce

$$\text{Model 3:} \quad M\frac{d^2X}{dt^2} = -KX - K_1V - K_2V^3, \quad \text{(B2.3)}$$

where we have added yet another parameter, which now relates the friction force to the cube of the velocity. The solution of Equation B2.3 requires advanced methods and is usually not treated in introductory courses. Note, however, that these three models are "nested": we obtain model 2 or model 1 from model 3 by setting certain parameters equal to 0.

Thus, with the single hypothesis $F = Ma$, we have at least three different models and could confront these models with the observations. Surely we believe that none of these is "correct," but that they are increasingly better descriptions of reality within the hypothesis.

Now suppose that the mass is a ball containing sand and that there is a hole in the bottom so that the sand falls out as the oscillations occur. In this case, our hypothesis is no longer correct, because $F = Ma$ assumes that the mass is constant. In more advanced physics courses, one learns that the appropriate hypothesis for the case in which the mass is a changing function of time, $M(t)$, is $F = (d/dt)$ (momentum), where momentum $= M(t)V$. This is an alternate hypothesis, which requires another series of models like the ones we just discussed.

when a better one arises (that is, don't fall in love with your model). Complicated models with more parameters and mechanisms will usually give better fits to data than simpler models, but if our models are as complicated as nature itself, then we may as well not bother with the model and focus only on the natural situation. Simpler models often provide insight that is more valuable and influential in guiding thought than accurate numerical fits. In fact, although the output of most models is numerical, the most influential models are the ones in which the numerical output is not needed to guide the qualitative understanding.

In summary, models allow us to tie together different bodies of data and aid in the identification of salient, necessary, and sufficient features of a system. The use of models while planning an experiment may help identify variables that will be confounded in the analysis of the results. Finally, models allow us to explore the parameter space and analyze multidimensional systems in ways that are virtually impossible from a purely empirical perspective.

Recognition of the model as a scientific tool has a number of important implications. First, one must try to validate assumptions before starting, or at least keep track of the untested assumptions. For example, the generally rancorous discussion concerning optimality theory in biology over the last twenty years was caused, in no small part, because both sides failed to recognize the nature of the assumptions and failed to clearly identify what was being tested and what was not being tested (e.g., Stephens and Krebs 1986; Mitchell and Valone 1990; Orzack 1993; Orzack and Sober 1994). The typical scenario often went like this: A model of an "optimally foraging animal" was constructed and compared with data. The data and model never matched completely, so opponents claimed that "optimal foraging" was disproved, while proponents modified the model and tried again to obtain agreement between the model and the data. And the argument continues.

30

The idea that models should be used as a principal tool in confronting hypotheses with data as arbitrator leads into a natural discussion of "model validation." It is a long-held and common view that in ecological studies, models should be "validated" by some kind of comparison of predictions of the model and the data that motivated it (e.g., Naylor and Finger 1967; Mankin et al. 1975; Shaeffer 1980; Leggett and Williams 1981; Feldman et al. 1984; Santer and Wigley 1990; Wigley and Santer 1990). Adopting a Popperian view, if the model is inconsistent with any of the data, then it (and the associated hypothesis) should be rejected. The model would be tested repeatedly, subjecting it to new challenges in the form of new empirical data. A model that withstood repeated challenges could be considered as "valid" only in the sense that it was not rejected. In contrast, adopting the Lakatosian view, all models will be found inconsistent with some of the data, and the question is which models are most consistent and which ones meet the challenges of new experiments and new data better. Thus, models are not validated; alternative models are options with different degrees of belief (see Oreskes et al. 1994 for an excellent discussion of this topic for models in the earth sciences). If one model clearly fits the existing data best and has proven ability to explain new data, we might have a very high degree of belief. It is not validated—but is better than the competitors. The favorite model of the current moment will likely be replaced by another model in the future. Levins (1966, 430–31) wonderfully states the situation:

A mathematical model is neither an hypothesis nor a theory. Unlike scientific hypotheses, a model is not verifiable directly by an experiment. For all models are both true and false. . . . The validation of a model is not that it is "true" but that it generates good testable hypotheses relevant to important problems. A model may be discarded in favor of a more powerful one, but it usually is simply outgrown when the live issues are not any longer those for which it was designed. . . . The multiplicity of models is imposed by the contradictory demands of a complex, heterogeneous nature and a mind that can only cope with a few variables at a time . . . individual models, while they are essential for understanding reality, should not be confused with that reality itself.

## TYPES AND USES OF MODELS

The ecological literature is filled with different kinds of models, which can be used for different kinds of investigations (Loehle 1983). One way to classify models is according to dichotomies. Here we specify some of these differences, and in the applications chapters you will see the different kinds of models in action.

### Deterministic and Stochastic Models

Deterministic models have no components that are inherently uncertain, i.e., no parameters in the model are characterized by probability distributions. In stochastic models, on the other hand, some of the parameters are uncertain and characterized by probability distributions. For fixed starting values, a deterministic model will always produce the same results, but the stochastic model will produce many different results depending on the actual values the random variables take.

### Statistical and Scientific Models

A scientific model begins with a description of how nature might work, and proceeds from this description to a set of predictions relating the independent and dependent variables. A statistical model foregoes any attempt to explain why the variables interact the way they do, and simply attempts to describe the relationship, with the assumption that the relationship extends past the measured values. Regression models are the standard form of such descriptions, and Peters (1991) argued that the only predictive models in ecology should be statistical ones; we consider this an overly narrow viewpoint.

## Static and Dynamic Models

Static models predict a response to input variables that does not change over time. Dynamic models involve responses that change over time. In this regard, dynamic models become more complicated because they often involve the link of the response between one period and the next.

## Quantitative and Qualitative Models

Quantitative models lead to detailed, numerical predictions about responses, whereas qualitative models lead to general descriptions about the responses. The ideal use of models is to develop quantitative models from which qualitative insights can be gained. It is often reasonable to test quantitative predictions that are based on simple models, using estimated or averaged parameters, with the intention of assessing how well the simple description of nature works. Qualitative models, on the other hand, can be used more broadly to describe regions in which one response is expected and regions in which a different response is expected. For example, when studying whether an insect of a given age and physiological state will oviposit on a host of a specified type, we might use a model (Mangel 1987) to divide the "age/state" plane into one region in which oviposition will occur and one in which it will not occur. A quantitative model would attempt to determine the precise location of the boundary, whereas a qualitative model would recognize that such a boundary exists and then ask how the responses would change in response to other parameters. Such predictions are quite testable (Roitberg et al. 1992, 1993).

33

*Models for Understanding, Prediction, and Decision*

We must recognize that in addition to different kinds of models there are different uses of models. We may model a natural system to broadly test our understanding of the mechanisms in the system. However, models usually lead to numerical predictions. In that case, we want to abstract qualitative, intuitive understanding from the broad pattern of the numerical predictions.

A model may be used for purposes of prediction. Such predictions can be both qualitative (e.g., "the system will/will not respond to this effect") and quantitative (e.g., "the level of the response will be . . ."). A model is most effective, of course, if it provides both understanding (of known patterns) and prediction (about situations not yet encountered).

Finally, we can use the model as part of a decision-making process. In this case, the model provides a means for evaluating the potential effects of various kinds of decisions. It is in this realm that models have the most to offer in terms of practical application, but also where the greatest potential danger lies.

## NESTED MODELS

Very often, we want to develop different models for the description of the same phenomenon. A particularly useful way of doing this is by adding complexity so that the "next model" contains the "previous model" as a special case, usually when some parameter (or parameters) is fixed. A family of models is called nested if the simpler models are special cases of the more complex models (see McCullagh and Nelder 1989 for a general discussion).

As a specific example, suppose we had a set of observations of population abundance $Y$ at a series of spatial sites, indexed by $i$, and a number of independent variables measured at the same sites, such as water availability, ground

34

cover, tree cover, insect abundance, etc. We denote these variables with $X_{i1}$, $X_{i2}$, $X_{i3}$, etc. (where $X_{ij}$ is the value of the $j^{th}$ measured variable at site $i$). One model relating these variables is

$$\log(Y_i) = p_0 + p_1 X_{i1} + p_2 X_{i2} + p_3 X_{i3} + E_i, \qquad (2.2)$$

where the $E_i$ represents a source of uncertainty and the parameters $p_i$ are determined during the confrontation with the data. A model such as Equation 2.2 is called a log-linear model, because the logarithm of the dependent variable $Y_i$ is assumed to be a linear function of the independent variables $\{X_i\}$. The model Equation 2.2 is one of a family of models that includes

$$\log(Y_i) = p_0 + p_1 X_{i1} + p_2 X_{i2} + E_i,$$

$$\log(Y_i) = p_0 + p_1 X_{i1} + E_i,$$

$$\log(Y_i) = p_0 + p_1 X_{i1} + p_3 X_{i3} + E_i,$$

$$\log(Y_i) = p_0 + p_2 X_{i2} + p_3 X_{i3} + E_i,$$

$$\log(Y_i) = p_0 + p_2 X_{i2} + E_i,$$

$$\log(Y_i) = p_0 + p_3 X_{i3} + E_i, \qquad (2.3)$$

as some of the special cases. All the models in Equation (2.3) are special cases of the full model when different parameters are set to zero; this family of models is said to be nested. The same is true for some of the models in Equation 2.1 (reader, which ones?).

Many ecological models can be treated as nested models. The Leslie life history model (Caswell 1989), used frequently for age- or size-structured populations, is

$$N_{a+1,t+1} = s_a N_{a,t}, \qquad \text{for } a > 1,$$
$$N_{1,t} = \sum_a m_a N_{a,t} \qquad (2.4)$$

where $N_{a,t}$ is the number of animals of age $a$ at time $t$, $s_a$ is the fraction of animals of age $a$ surviving to age $a + 1$, and $m_a$ is the reproduction by animals of age $a$. A special, and therefore nested, case of the Leslie model is one without age structure, which can be obtained by assuming that the survival at each age is the same (so set $s_a = s$) and that reproduction at each age is the same (so set $m_a = m$). Then if $N_t$ is the total population size and $B_t = mN_t$,

$$N_{t+1} = sN_t + B_t. \tag{2.5}$$

The alternative to nested models is to consider models that are structurally different, where we cannot change a parameter to obtain one model from the other. In dealing with non-nested models we can no longer simply ask if we obtain better fits to the data by making the model more complex, but we must see how well the alternative models fit the data.

## MODEL COMPLEXITY

Perhaps the most difficult decision in model building is "How complex should the model be?" With microcomputers and modern software it is easy to build models quickly, to run the models, and generate lots of output. It takes only a few minutes to add additional variables to the model and if we continue for a few hours, we could have a model with dozens or hundreds of variables. What is the best-sized model? There are usually two major factors influencing the answer to this question. On one hand, we can always imagine that the model would be better ("more realistic") if we added another component to it—something we have observed in nature and hate to leave out. On the other hand, if we have a smaller model, the computer will run faster, fewer parameters will be needed, and the output will be easier to understand. Most neophytes are tempted to build very large models, and we urge you to resist this temptation. Of course, the best-sized model depends on the purpose of the model. Given this objective, the basic rule about model size is

# Let the data tell you.

There are quantitative methods for determining the optimal size of a particular model (Ludwig and Walters 1985; Linhart and Zucchini 1986; Walters 1986; Punt 1988; Gauch 1993). If the model is too simple, we risk leaving out significant components of the system. If the model is too complex, we will not have sufficient information in the data to distinguish between the possible parameter values of the model.

For example, many ecological analyses of population dynamics rely on the Leslie matrix with age-specific survival and fecundity. If we wish to make projections of the population size and have estimated survival and fecundity for only a few individuals, we have the choice of several models. The simplest model (e.g., Equation 2.5) would average the survival and fecundity over all ages; the most complex model (e.g., Equation 2.4) would estimate the survival and fecundity at each age from the data. If the species is long lived and the number of individuals for whom survival and fecundities has been measured is small, estimates of the age-specific survival and fecundity are likely to be poor, and it would be better either to use a single value for all ages or at least to average survival and fecundity over age groups. The number of ages aggregated should depend on the amount of data available and the number of age classes considered.

Linhart and Zucchini (1986) provide a formal framework for considering different levels of model complexity in the reliability of model predictions. Their approach distinguishes between prediction error due to approximation, which decreases as model complexity increases, and prediction error due to estimation, which increases as model complexity increases. For any model and amount of data, the total prediction error will decrease and then increase as model complexity increases—with respect to reliability of prediction, there is an optimal level of model complexity.

Linhart and Zucchini's approach is consistent with almost all quantitative work in this area that suggested the optimal model size is much smaller than intuition dictates. Ludwig and Walters (1985) obtained better predictions about management actions from a non-age-structured model, even when the data were derived, by simulation, from an age-structured model. That is, the "wrong" model can do better than the "right" model in prediction if parameters must be estimated. Similarly, Punt (1988) found very simple models of fisheries management, which often ignored substantial amounts of data, outperformed more complex models when parameters had to be estimated and decisions made.

When the objective is something other than prediction accuracy, the complexity of the optimal mode may be quite different. In Chapter 10, we show a fisheries example where a complex model fits the available data no better than a simpler model. However, the uncertainty in the sustainable harvest is quite low for the simple model, but high for the complex model. In this case the simple model under-represents the uncertainty, and we believe that a more complex model provides a better representation of the uncertainty.

The complexity of the optimal model will depend on the use of the model and on the data. Part of the work of the ecological detective is to iterate between alternative models, to understand their strengths and weaknesses, and to recognize that the most appropriate model will change from application to application.

# Probability and Probability Models: Know Your Data

## DESCRIPTIONS OF RANDOMNESS

The data we encounter in ecological settings involve different kinds of randomness. Many ecological models describe only the average, or modal, value of a parameter, but when we compare models to data, we need methods for determining the probability of individual observations, given a specific model and a value for the mean or mode of the parameter. This requires that we describe the randomness in the data. Similarly, when we build a model and want to generate a distribution of some characteristic, we first need a way to quantify the probability distribution associated with this characteristic. This involves understanding both the nature of your data and the appropriate probabilistic descriptions.

We assume that readers of this book are familiar with the normal or Gaussian distribution (the familiar "bell-shaped curve"). However, many of the distributions in nature are not normal. The purpose of this chapter is to introduce ideas about probability, describe a wide range of useful probability distributions (and consider biological processes that give rise to these distributions), and provide you with the tools you need to use these distributions in your work. We begin with advice on data and then review the concepts of probability. After that, we describe a number of different probability distributions and some of their ecological applications. We close with a description and illustration of the "Monte Carlo" method for generating data and testing models.

A modest university library will have fifty to one hundred textbooks on probability that cover the material we treat here in more detail. So why do we bother? There are two main reasons. First, we want to motivate you to be interested in other than normal distributions. Second, we want to provide enough detail so that when the distributions are used in subsequent applications, the book is self-contained. We suggest that you skim the distributional information now and return to it as needed in later chapters.

## ALWAYS PLOT YOUR DATA

Ecological systems are complex. For this reason, we can hope to observe only a very small fraction of the possible variables. The largest field research programs barely scratch the surface of what could be measured. Indeed, the key questions in the design of ecological research are what experiments to perform, what to measure, and how to measure it. Whole new avenues of research have been developed based on new measurement methodologies such as radiotracking, starch gel electrophoresis, DNA fingerprinting, and individual identification of animals by natural marks.

When confronting alternative models with data, we must decide not only which models, but also which data to use. In practice we often observe more than one feature of the ecological system. For example, population surveys may be conducted in many different years, and these surveys provide the major source of information for the model. However, in some years there may be additional direct measurements of birth or death rates.

So what is the first step? Plot your data. Get to know them by using standard computer graphic routines to fit various curves (linear, polynomial, logarithmic, exponential). When there are more than two variables, plot the data in many ways and look for correlation. Think about plausible functional relationships.

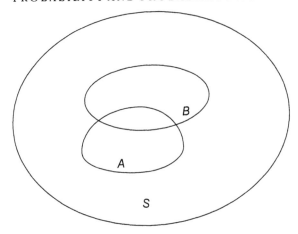

FIGURE 3.1. The probability of the event $A$ is the area of $A$, however area might be defined, divided by the area of $S$, which is the collection of all possible outcomes of the experiment.

## EXPERIMENTS, EVENTS, AND PROBABILITY

In probability theory, we are concerned with the occurrence of "events" that can be thought of as outcomes of experiments. The probability of an event $A$ is denoted by

$$\Pr\{A\} = \text{probability that the event } A \text{ occurs.} \qquad (3.1)$$

It is helpful to think of probability in the following way. First, we imagine all the possible outcomes of the experiment and call this collection of outcomes $S$. A smaller collection of outcomes, $A$, has probability defined as the "area" of $A$ divided by the "area" of $S$, with "area" suitably defined (Figure 3.1). Particular probability models give different definitions of what "area of $A$" really means. In any case,

$$\Pr\{A\} = \text{probability that the event } A \text{ occurs}$$

$$= (\text{area of } A)/(\text{area of } S). \qquad (3.2)$$

Continuing to use this figure and the definition of probability in Equation 3.2, we see that the probability that one of two events $A$ or $B$ occurs is

41

$$\Pr\{A \text{ or } B\} = \Pr\{A\} + \Pr\{B\} - \Pr\{A \text{ and } B\}. \qquad (3.3)$$

In the future, we will use $\Pr\{A,B\}$ for the probability that both $A$ and $B$ occur.

### Conditional Probability

Referring again to Figure 3.1, suppose that we know that event $A$ occurred. What is the probability that $B$ occurred, given the knowledge about $A$? This kind of question arises all the time in ecological detection as we use models to make predictions about data and data to make inferences about different models.

If $A$ occurred, then the collection of all possible outcomes of the experiment is no longer $S$, but must be $A$. From the definition Equation 3.2,

$\Pr\{B$ occurred, given that $A$ occurred$\}$

$\qquad =$ (area common to $A$ and $B$)/(area of $A$). $\qquad (3.4)$

We use $\Pr\{B|A\}$ to denote the probability that $B$ occurs given that $A$ occurs. Dividing the numerator and denominator of the right-hand side of Equation 3.4 by the area of $S$ and using the new notation, we have

$$\Pr\{B|A\} = \Pr\{A,B\}/\Pr\{A\}. \qquad (3.5)$$

By analogy, since $A$ and $B$ are fully interchangeable here, we must also have

$$\Pr\{A|B\} = \Pr\{A,B\}/\Pr\{B\}. \qquad (3.6)$$

We define two events as independent if knowing that one of them occurred does nothing to change our idea about the probability of the other one occurring. Thus, if $A$ and $B$ are independent,

$$\Pr\{A|B\} = \Pr\{A\} \quad \text{and} \quad \Pr\{B|A\} = \Pr\{B\}. \qquad (3.7)$$

Using these in either Equation 3.5 or Equation 3.6, we see that for independent events

$$\Pr\{A,B\} = \Pr\{A\} \Pr\{B\}. \qquad (3.8)$$

Equation 3.8 is often given as the definition of independent events, but it is actually derived from the definition based on conditioning.

### Bayes' Theorem

The challenge in ecological detection (and all statistical science, for that matter) is to determine how to use the information contained in data and Bayes' theorem is a very powerful method.

From Equation 3.6, we see that $\Pr\{A,B\} = \Pr\{A|B\}\Pr\{B\}$. Using this in Equation 3.5, we have

$$\Pr\{B|A\} = \Pr\{A,B\}/\Pr\{A\} = \Pr\{A|B\}\,\Pr\{B\}/\,\Pr\{A\}. \quad (3.9)$$

The extreme left- and right-hand sides of this formula are called Bayes' theorem. It is most handy when there are a number of possible but mutually exclusive outcomes $B_1, B_2,$ $\ldots, B_N$, one of which must occur when $A$ occurs. The natural generalization of Equation 3.9 is to ask for the probability that $B_i$ occurs given that $A$ occurs (Figure 3.2). Following the reasoning that led to Equation 3.9, you should show that

$$\Pr\{B_i|A\} = \Pr\{A|B_i\}\,\Pr\{B_i\} \left/ \sum_{j=1}^{N} \Pr\{A|B_j\}\Pr\{B_j\}. \right. \quad (3.10)$$

Two hints: note that (1) the numerator on the right-hand side is the joint probability $A$ and $B_i$, and (2) the denominator is the same as $\sum_{j=1}^{N} \Pr\{A,B_j\}$. What must be true about this expression?

We now illustrate some of the nuances of conditional probability with two examples (Bar-Hillel and Falk 1982).

*Predator and Prey.* Imagine a rabbit wandering through the forest. If it comes within a critical distance of a predator (e.g., a fox or coyote), there is a probability $P_A$ that the predator will attack. In addition, suppose that the rabbit often does not observe the predator directly, but uses various

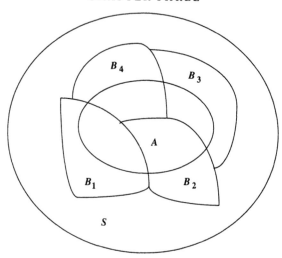

FIGURE 3.2. An illustration of Bayes' theorem for a case in which, when event $A$ occurs, one of four other possible events $B_1, \ldots, B_4$ may also occur.

cues (e.g., scent) of the predator's presence. Assume that $P_s$ is the probability that if the rabbit obtains such a signal, the predator is within the critical attack distance. Once the rabbit obtains such a signal, what is the probability of an attack? The answer is not $P_s P_A$, as tempting as it may seem.

In order to answer the question, we introduce events:

$A$ = event of being attacked,
$P$ = event of predator present within the critical attack distance,
$S$ = event of receiving the cue, $\qquad$ (3.11)

so that the data are

$$\Pr\{A|P\} = P_A,$$

$$\Pr\{P|S\} = P_s. \qquad (3.12)$$

The probability we wish to calculate is

$$\text{Pr}\{\text{attack, given the signal}\} = \text{Pr}\{A|S\}$$
$$= \text{Pr}\{A,S\}/\text{Pr}\{S\}. \qquad (3.13)$$

Applying Bayes' theorem,

$$\text{Pr}\{A|S\} = \text{Pr}\{A,S\}/\text{Pr}\{S\} = \text{Pr}\{S|A\}\,\text{Pr}\{A\}/\text{Pr}\{S\}. \qquad (3.14)$$

The key piece of information in this equation is $\text{Pr}\{S|A\}$, the probability that a signal is obtained when an attack actually does occur. This is not available from the given data and (particularly if the predator is smart) could, in fact, be 0! Thus, if the rabbit is a careless Bayesian, it may misjudge the meaning of a cue.

*Smith's Children* (Bar-Hillel and Falk, 1982). Smith has two children. You meet Smith and a child who is a boy. What is the probability that the other child is also a boy?

There are two lines of reasoning about this problem. If the sexes of the children are determined independently and with equal probability, then by independence

Pr{second child is a boy | first child is a boy}
$$= \text{Pr}\{\text{second child is a boy}\} = 1/2. \qquad (3.15)$$

The second line of reasoning is the following. Before meeting the first child, the possible events in Smith's family are $\{GG, GB, BG, BB\}$, where $G$ denotes girl and $B$ denotes boy. The information that the child we met is a boy eliminates $GG$ as one of the possible events, so that given this information, the possible events are $\{GB, BG, BB\}$. With this line of reasoning, if each family mix is equally likely, the probability that the second child is a boy is $1/3$.

Clearly, these two lines of reasoning cannot be correct. One approach is to forget about the problem, since "Both arguments appear reasonable and both have been used in practice. What to do about the contradiction? The easiest way out is that of a formalist, who refuses to see a problem if it is not formulated in an impeccable manner. *But problems are not solved by ignoring them.*" (Feller 1971, 12, emphasis added.)

The difficulty lies in how we use the information that one of the children is a boy. We want to find

Pr{family type is $BB$ | met child is a boy}
= Pr{family type is $BB$, met child is a boy}/
Pr{met child is a boy}.   (3.16)

Allowing all four possible family types, we have:

| Family type | Prior probability | Pr{meeting a boy, given family type} |
|---|---|---|
| $BB$ | 1/4 | 1 |
| $BG$ | 1/4 | 1/2 |
| $GB$ | 1/4 | 1/2 |
| $GG$ | 1/4 | 0 |

Assuming independence of the met child and the family type, the joint probability of family type and meeting a boy is

Pr{family type is $BB$, met child is a boy}
$$= (1/4) \times 1 = 1/4,$$
Pr{family type is $BG$, met child is a boy}
$$= (1/4) \times (1/2) = 1/8,$$
Pr{family type is $GB$, met child is a boy}
$$= (1/4) \times (1/2) = 1/8,$$
Pr{family type is $GG$, met child is a boy}
$$= (1/4) \times 0 = 0,$$

so that we have

Pr{met child is a boy} $= 1/4 + 1/8 + 1/8 = 1/2$,

and using this in Equation 3.16 we conclude that

Pr{second child is a boy | met child is a boy} $= 1/2$.   (3.17)

Thus, the first line of reasoning is correct and the second is not. We encourage you to think about what was wrong with the second line of reasoning. In particular, does the fact of meeting a boy change the probabilities for the four family types?

*Random Variables, Distribution Functions, and Density Functions*

A random variable $Z$ is one that can take more than one value in which the values are determined by probabilities. If the random variable takes discrete values, we write

$$\Pr\{Z = k\} = f_k, \tag{3.18}$$

where $0 \leq f_k \leq 1$ and $\Sigma_k f_k = 1$. For example, $Z$ might take the values $1, 2, \ldots 10$, each with equal probability $0.1$. Then $f_k = 0.1$ and $\Sigma_{k=1}^{z} f_k$ is the probability that $Z \leq z$, which we shall denote by $F(z)$; Figure 3.3 illustrates this idea. $F(z)$ is called the cumulative distribution function. Cumulative distribution functions should have the following properties: (i) as $z \to -\infty$, $F(z) \to 0$; (ii) as $z \to \infty$, $F(z) \to 1$; (iii) $F(z)$ never decreases as $z$ increases.

When the data are continuous variables, such as lengths, weights, or time, we cannot write the probability distributions in the same way since $z$ can take an infinite number of values in any finite interval. In such a case, we begin with the cumulative distribution function, also indicated by $F(z)$ and which has the same interpretation,

$$F(z) = \Pr\{Z \leq z\}. \tag{3.19}$$

An example of such a cumulative distribution function is

$$F(z) = \begin{cases} 0 & \text{if } z < 0, \\ 1 - e^{-rz} & \text{if } z \geq 0, \end{cases} \tag{3.20}$$

which is called the "negative exponential distribution function" (Figure 3.4).

When $Z$ is continuous, we can no longer speak of the event "$Z = z$." Instead, we consider the chance that $Z$ takes a value in a small neighborhood $\Delta z$ of $z$ and we can evaluate it with the following logic (we encourage you to sketch out this idea using Figure 3.4):

47

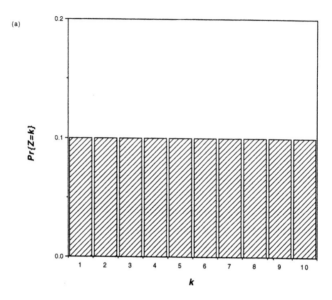

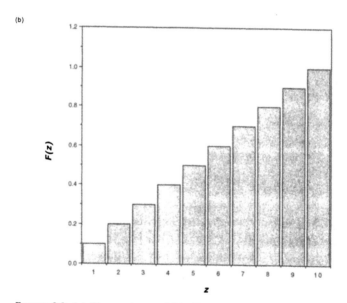

FIGURE 3.3. (a) The random variable $Z$ may be any of $1, 2, \ldots, 10$, each with equal probability $f_k = 0.1$. Such a random variable is said to be uniformly distributed. (b) The probability that $Z$ is less than or equal to $z$, $F(z)$, is obtained by summing the $f_k$.

$\Pr\{z \le Z \le z + \Delta z\}$
$$= \Pr\{ Z \le z + \Delta z\} - \Pr\{Z \le z\}$$
$$= F(z + \Delta z) - F(z). \tag{3.21}$$

Since $\Delta z$ is assumed to be a small value, we use a Taylor expansion[1] of $F(z + \Delta z)$

$F(z + \Delta z)$
$$= F(z) + F'(z)\Delta z + \frac{1}{2}F''(z)\ (\Delta z)^2 + \dots \tag{3.22}$$

We scoop all the terms involving high powers of $dz$ into the single expression $o(\Delta z)$. This handy notation will be used in other places in the book. Equation 3.22 becomes

$$F(z + \Delta z) = F(z) + F'(z)\Delta z + o(\Delta z), \tag{3.23}$$

and using this in Equation 3.21,

$$\Pr\{z \le Z \le z + \Delta z\} = F'(z)\Delta z + o(\Delta z). \tag{3.24}$$

The derivative $F'(z)$ is called the probability density function and is denoted by the symbol $f(z)$. For example, a continuous distribution might be used to represent the lengths of animals in a population. When such a graph is drawn using real data, it is often a histogram, where the ordinate is the number of individuals falling in each length interval. When it is represented as a continuous curve, the appropriate label is $f(z)$, which is interpreted as the frequency distribution of outcomes. For the negative exponential distribution function, the probability density function is (Figure 3.4b)

$$f(z) = re^{-rz}. \tag{3.25}$$

These ideas of probability can be nicely illustrated by a study of predation (Box 3.2).

[1]You are going to need six facts from calculus in order to completely understand this chapter. They are given in Box 3.1.

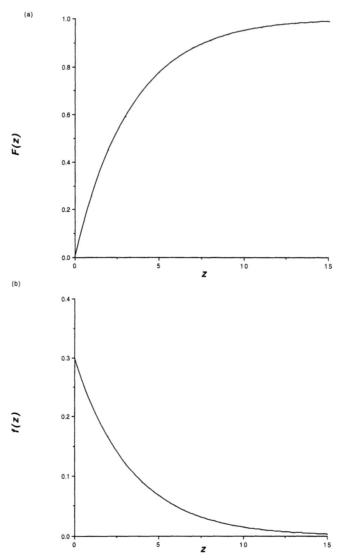

FIGURE 3.4. (a) The negative exponential distribution function $F(z) = 1 - e^{-rz}$ for $r = 0.3$. (b) The corresponding density function $f(z) = re^{-rz}$.

BOX 3.1

THE CALCULUS FACTS YOU NEED FOR THIS CHAPTER

1. Definition of the derivative:

$$\frac{dF}{dx} = F'(x) = \lim_{\Delta x \to 0} \frac{F(x + \Delta x) - F(x)}{\Delta x}.$$

2. Derivative of the exponential function:

$$\frac{d}{dx} e^{kx} = k e^{kx}.$$

3. Exponential function as a limit:

$$\lim_{n \to \infty} \left( 1 + \frac{x}{n} \right)^n = e^x.$$

4. Integral as a limit of a sum:

$$\int_a^b h(z)\, dz = \lim_{\Delta z \to 0} \sum_{z=a}^{b} h(z)\, \Delta z,$$

where the summation goes from $z = a$ to $z = b$ in steps of $\Delta z$.

5. Taylor expansion for a function of one variable:

$$F(x) = F(a) + F'(a)(x - a) + \frac{1}{2} F''(a)(x - a)^2 + \ldots,$$

where $F'(a)$ is the first derivative of $F(x)$ evaluated at $x = a$, $F''(a)$ is the second derivative of $F(x)$ evaluated at $x = a$ and "+ ..." means terms that are higher powers of $(x - a)$, such as $(x - a)^3$, $(x - a)^4$, etc.

BOX 3.1 CONT.

Taylor expansion for a function of two variables:

$$F(x,y) = F(a,b) + \frac{\partial F(a,b)}{\partial x}(x-a) + \frac{\partial F(a,b)}{\partial y}(y-b)$$
$$+ \frac{1}{2}\left( \frac{\partial^2 F(a,b)}{\partial x^2}(x-a)^2 \right.$$
$$+ 2\frac{\partial^2 F(a,b)}{\partial x\,\partial y}(x-a)(y-b)$$
$$\left. + \frac{\partial^2 F(a,b)}{\partial y^2}(y-b)^2 \right),$$

where $\partial F(a,b)/\partial x$ and $\partial F(a,b)/\partial y$ are the first partial derivatives of $F(x,y)$ with respect to $x$ and $y$, evaluated at $x = a$ and $y = b$; $\partial^2 F(a,b)/\partial x^2$, $\partial^2 F(a,b)/\partial x\,\partial y$, and $\partial^2 F(a,b)/\partial y^2$ are the second partial derivatives with respect to $x$, with respect to $x$ once and $y$ once, and $y$, evaluated at $x = a$ and $y = b$.

6. The chain rule:

$$\frac{d}{dx}f(g(x)) = f'(g)g'(x).$$

*Expectation, Variance, Standard Deviation, and Coefficient of Variation*

We denote average, mean, or expectation by $E\{\ \}$. For a discrete random variable and for any function $g(z)$, we define the expectation by

$$E\{Z\} = \sum_z z f_z \quad \text{or} \quad E\{Z\} = \int z f(z)\, dz \quad (3.26)$$

for discrete and continuous random variables, respectively. (Refer again to Box 3.1, for the calculus facts regarding the relationship between sums and integrals.)

BOX 3.2

Random Search and Predation

The rules of probability we just discussed provide some interesting insights into predation. We encourage you to work out all the details of this example, because it will help solidify the notions of probability and the notation we use.

Suppose that an organism searches for food and

Pr{finding food in the next increment of time $\Delta t$ | no food
found thus far} $= c\Delta t + o(\Delta t)$,

where $c$ is a fixed constant (this will turn out to be very important) and $o(\Delta t)$ represents terms that are higher powers of $\Delta t$. We set

$$Q(t) = \text{Pr\{not finding food in the interval } [0, t]\}$$

and note that for the animal not to find food in the interval $[0, t + \Delta t]$ it first must not find food in the interval $[0, t]$ and then not find food in the next $\Delta t$. Assuming that these are independent events (what is the biological implication of this assumption?),

$$Q(t + dt) = Q(t)[1 - c\Delta t + o(\Delta t)].$$

Subtracting $Q(t)$ from both sides we have

$$Q(t + \Delta t) - Q(t) = -cQ(t)\,\Delta t + o(\Delta t).$$

Dividing both sides by $dt$ and letting $\Delta t \to 0$ gives the derivative of $Q_0(t)$ on the left-hand side (see Box 3.1). Since $o(\Delta t)$ denotes terms that are like $(\Delta t)^2$, etc., $o(\Delta t)/\Delta t \to 0$ as $\Delta t \to 0$. Thus, the difference equation becomes a differential equation for $Q(t)$:

BOX 3.2 CONT.

$$\frac{dQ}{dt} = - c\, Q(t).$$

We see that the derivative of $Q_0(t)$ is a constant times $Q(t)$. This means that $Q(t)$ must be an exponential (see Box 3.1) of the form $Q(t) = Ae^{-ct}$. Since $Q(0) = 1$ (no food is found before the start of the search for food), the constant $A = 1$. We have demonstrated that

$$\Pr\{\text{not finding food in } [0,t]\} = Q(t) = e^{-ct}.$$

Koopman (1980) derives this formula in a different way in which the biological interpretation of $c$ becomes more apparent. Suppose that the search takes place in a "large" region of area $\mathcal{A}$ that contains the food item. Assume that $W$ is the detection width of the searching animal, in the sense that if the food is within a distance $W/2$ of the animal, the food is discovered. If $v$ is the speed of the searching animal, in the interval of time $dt$ the animal covers an area $Wv\Delta t$ and detects the food with probability $Wv\Delta t/\mathcal{A}$. Envision the time interval $[0,t]$ divided into $n$ legs of length $t/n$, so that $\Delta t = 1/n$. Assuming that detection on each leg is independent of previous legs gives

$\Pr\{\text{no detection of food in } [0,t]\}$
$$= [\Pr\{\text{no detection of food on a single leg}\}]^n$$
$$= \left( 1 - \frac{Wvt}{\mathcal{A}n} \right)^n.$$

In the limit (see Box 3.1) that $n \to \infty$, the right-hand side of this expression becomes $e^{-Wvt/\mathcal{A}}$:

$$\Pr\{\text{no detection of food in } [0,t]\} = e^{-Wvt/\mathcal{A}},$$

## BOX 3.2 CONT.

so that the interpretation of $c$ is $c$ = detection rate = $Wv/\mathcal{A}$, and these parameters—$W$, $v$, and $\mathcal{A}$—can be measured independently of the searching process. Because $Q(t) = e^{-ct}$ and it is only possible to take the exponential of dimensionless quantities, we conclude that the units of $c$ (which are often denoted by $[c]$) must be 1/time. Since the units of $W$ are length, of $v$ are length/time and of $\mathcal{A}$ are (length)$^2$, we see that $Wv/\mathcal{A}$ has units of 1/time, as it should if our analysis is correct.

We shall now use notions of conditional probability to demonstrate the "memoryless property" of this model, assuming once again that $c$ is a fixed and certain parameter. We begin with $Q(t) = e^{-ct}$ and ask: What is the probability that the animal does not find food between $t$ and $t + s$, given that it did not find food up to time $t$? Applying the definition of conditional probability,

Pr{no food in $(t, t + s)$ | no food in $(0,t)$}

$$= \frac{\text{Pr\{no food in } (t, t + s) \text{ and no food in } (0,t)\}}{\text{Pr\{no food in } (0,t)\}}.$$

Since the numerator is the same as no event in the interval from 0 to $t + s$, we have

Pr{no food in $(t, t + s)$ | no food in $(0,t)$}
$$= e^{-c(t+s)}/e^{-ct}$$
$$= e^{-cs} = \text{Pr\{no food in } (0,s)\}.$$

Thus, the fact that no food was found before time $t$ provides no information about the probability of events after time $t$. The predator in this model does not "learn." This is somewhat discomforting, because we expect that a failed search

> ### BOX 3.2 CONT.
>
> should provide information about the search rate $c$. But remember that we assumed $c$ to be known and fixed. Later in this chapter, after discussing the gamma density, we will consider how failed searches may change our view of the frequency distribution of $c$, if we allow it to be uncertain.

These definitions generalize for any function $g(Z)$; for example, $E\{g(Z)\} = \Sigma_z\ g(z)f_z$. The generalization is very handy for computing measures of variability about the average. If we denote the average by $m_1$, the variance of the random variable $Z$ is

$$\begin{aligned} \text{VAR}\{Z\} &= E\{(Z-m_1)^2\} \\ &= \sum_z (z-m_1)^2 f_z \quad \text{or} \quad \int (z-m_1)^2 f(z)\ dz, \quad (3.27) \end{aligned}$$

depending on whether the random variable is discrete or continuous. The variance gives a sense of the "spread" of values of $Z$ around the average.

Two other measures of variability of $Z$ are the standard deviation,

$$\text{SD}\{Z\} = \sqrt{\text{VAR}\{Z\}} \quad\quad (3.28)$$

and the coefficient of variation

$$\text{CV}\{Z\} = \frac{\text{SD}\{Z\}}{E\{Z\}} . \quad\quad (3.29)$$

We are partial to the coefficient of variation as a measure of variation for the following reason. The standard deviation has the same units as $Z$, so that the coefficient of variation is a dimensionless measure of variability in which the scaling is relative to the mean. To see why this kind of scaling is important, consider the following two sequences of numbers:

A: 45, 32, 12, 23, 26, 27, 39

B: 1040, 1027, 1007, 1018, 1021, 1022, 1034

When asked which sequence is more variable, most people will say that sequence A is more variable. Sequence B is sequence A plus 995, so that the variance of these two sequences is exactly the same. However, the coefficient of variation of sequence B is much smaller than that of A. You should (i) verify that this is true by computation, and (ii) understand the reason for this being true. Some cognitive psychologists have argued that this is a matter of context: "Which series exhibits more variability? Most people answer series A. However, the statistical measure of *variance*—which indicates the amount of irregular variations from the mean of a series of numbers—is *the same* for both series. Series B is simply series A plus a constant. However, intuitive judgments of variability are usually influenced by the size or context of the series or objects. That is subjectively relative variability is more salient than variability per se" (Hogarth 1980, 44).

But when numbers have units, both the magnitude and the variability have meaning. For example, suppose that we measure the weights of five rodents and these are 0.079, 0.120, 0.085, 0.099, and 0.100 kg respectively. The average weight is 0.0966 kg, the variance is $2.018 \times 10^{-4}$ $kg^2$ (why $kg^2$?), and the coefficient of variation is 0.147. If the animals were weighed in grams rather than kilograms, the average would be 96.6 g and the variance 201.84 $g^2$ but the coefficient of variation would remain the same at about 15%.

By using the coefficient of variation, one takes this comparison out of the realm of the subjective and into the realm of the objective, with a measure of variation that is context-free because it has no dimensions. There is a tradition in ecology, which we elaborate during the discussion of the Poisson distribution, of comparing the mean and variance of data in order to determine whether the subject of study is "clumped" or not. This can only make sense if the random variable $Z$ is dimensionless.

## The Delta Method

When $g(z)$ is nonlinear, $E\{g(Z)\}$ is generally not equal to $g(E\{Z\})$. We encourage you to try out a numerical investigation for $g(z) = z^4$ using both the numerical data in Figure 3.3 and the negative exponential distribution of Figure 3.4. Because $E\{g(Z)\}$ may be difficult to find, an approximation commonly used is the "delta method" (Seber 1980). As before, let $m_1 = E\{Z\}$ and construct a two-term Taylor expansion $g(Z)$ around $m_1$:

$$g(Z) = g(m_1) + g'(m_1) \, (Z - m_1)$$
$$+ \frac{1}{2} g''(m_1) \, (Z - m_1)^2 + \ldots, \qquad (3.30)$$

where, also as before, $g'(m_1)$ and $g''(m_1)$ denote the first and second derivatives of $g(z)$ evaluated at $z = m_1$. Taking the expectation and ignoring all the terms represented by the ellipsis "$+ \ldots$," we have

$$E\{g(Z)\} = E\{g(m_1)\} + E\{g'(m_1) \, (Z - m_1)\}$$
$$+ \frac{1}{2} E\{g''(m_1) \, (z - m_1)^2\}. \quad (3.31)$$

You should verify from the definition of expectation that for any constant $c$,

$$E\{c\} = c \quad \text{and} \quad E\{cg(Z)\} = cE\{g(Z)\} \qquad (3.32)$$

and that

$$E\{(Z - m_1)\} = 0. \qquad (3.33)$$

Since $g(m_1)$, $g'(m_1)$, and $g''(m_1)$ are constants, Equation 3.31 becomes

$$E\{g(Z)\} = g(m_1) + \frac{1}{2} g''(m_1) \, \mathrm{VAR}(Z). \qquad (3.34)$$

We prefer to call this the method of "navy math," since it was commonly used by scientists in the Operations Evaluation Group (OEG) (see Tidman 1984) during World War II (Morse 1977) as a quick means of computing expectations. Those scientists and the ones who followed (Mangel 1982) are the inspiration for the part played by Kelly McGillis in *Top Gun*.

## PROCESS AND OBSERVATION UNCERTAINTIES

Before discussing particular probability distributions, let us spend time thinking about how stochasticity enters into ecological models. Ecological models often begin with a description of the processes of interest (e.g., birth rates, death rates, migration rates, etc.). For this reason, these models are sometimes called "process models." Uncertainty may enter into these processes because parameters vary in unpredictable ways.

To collect data about an ecological system, we observe it, and there will usually be uncertainty associated with the observations. For instance, suppose that we model a population by

$$N_{t+1} = sN_t + b_t, \tag{3.35}$$

where $N_t$ is the number of animals in the population at the start of period $t$, $s$ is a survival probability from $t$ to $t + 1$, and $b_t$ is the number of new individuals added in the interval $t$ to $t + 1$.

Uncertainty could enter in a number of different ways. For example, if birth rates fluctuate from one year to the next, we could write

$$N_{t+1} = sN_t + b_t + W_t, \tag{3.36}$$

where $W_t$ represents "process uncertainty," "process stochasticity," "process error," or "process noise" (depending on the particular subfield of ecology, all these terms are used). We use upper case to remind ourselves that $W_t$ is drawn from a distribution; a particular value would be denoted by $w_t$. In

59

principle, $W_t$ could arise from a number of the distributions we describe below and could depend on population size.

Since it is likely that there is uncertainty associated with the observations, we describe the observation model as

$$N_{\text{obs},t} = N_t + V_t, \qquad (3.37)$$

where $N_{\text{obs},t}$ is the observed population size at time $t$ and the "observation uncertainty" (or any of the other terms) $V_t$ might also depend on population size.

The process and observation models are now combined into a "full" model of the system:

$$N_{t+1} = sN_t + b_t + W_t,$$

$$N_{\text{obs},t} = N_t + V_t. \qquad (3.38)$$

To complete the model, we must specify the distributions of $W_t$ and $V_t$ and the initial population size. We shall return to this model at the end of the chapter, once the requisite skills are developed.

Since ecological detection involves comparing different models, it is useful at this point to think about other versions of the observation model.

*Bias.* Field methods for estimating animal abundance usually involve an unknown bias. For example, not all animals may be seen. In air surveys of marine mammals there is usually an unknown proportion of the animals below the surface. Transect counts of birds or smaller mammals almost always involve a fraction of the animals that cannot be seen from the observer's platform. To account for this effect, we might modify the observation model to

$$N_{\text{obs},t} = qN_t + V_t. \qquad (3.39)$$

Here, the parameter $q$ allows for bias of the observation system: When $q$ is less than 1 we tend to undercount the animals, and when $q$ is greater than 1 we tend to overcount them. As before, $V_t$ represents the observation uncertainty. It is almost always helpful, and frequently essential, to do experiments to determine $q$. However, in some instances, as in fisheries, we must estimate $q$ from the same data that we use to estimate the parameters of the process model.

*Nonlinearity.* We generalize the observation model further by including a nonlinear relationship between true abundance and observed abundance:

$$N_{\text{obs},t} = q(N_t)^c + V_t. \tag{3.40}$$

When $c$ is greater than 1 the estimated abundance rises more rapidly than real abundance, and when $c$ is less than 1 the estimated abundance changes less than real abundance.

*A Detection Threshold.* There may be a minimum threshold population size below which no animals can be seen, such as species where some proportion of the population finds hiding places. In this case, the observation model becomes

$$N_{\text{obs},t} = \max\{a + q(N_t)^c + V_t, 0\}, \tag{3.41}$$

where $\max\{A,B\} = A$ if $A > B$ and $\max\{A,B\} = B$ otherwise. If $A < 0$, it represents the population density below which no animals can be seen. If $A > 0$, some animals will appear to be present even when none are present. This could be due, for example, to improper species identification.

In summary, there is always an observation process interposed between the ecological system and our notebooks. Every effort should be made to understand, calibrate, and model the observation process. Doing this is an essential component of ecological detection.

*Additional Data.* In some years we may have additional sources of data. For example, suppose that in one year we had also conducted a study that provided an estimate of the number of deaths, in addition to the annual survey of abundance. Our model predicts the number of deaths as

$$D_t = (1 - s)N_t, \tag{3.42}$$

where $D_t$ is the number of deaths in year $t$. If we assume that the process uncertainty is entirely due to variation in births, then the observation model for deaths is

$$D_{\text{obs},t} = (1 - s)N_t + V_d, \qquad (3.43)$$

where $V_d$ is the uncertainty associated with the observation of the number of dead animals. Our model now predicts both the number of animals and the number of deaths, and when we see how well alternative models fit the data, we can compare the predictions with these observations. In later chapters, we will explore how to use multiple observations in a more rigorous framework.

However, we cannot conduct ecological detection without knowledge of the probability distributions that might describe the various kinds of uncertainty. This is what we consider next.

## SOME USEFUL PROBABILITY DISTRIBUTIONS

We now provide a review of a number of probability distributions that are tools for the ecological detective. We encourage you to skim this section now and return to it as the distributions are used in subsequent chapters. *However, whether or not you read it carefully now, you should read the next section on the Monte Carlo method.*

This review is not comprehensive. Our goal is to provide enough information so that you will know how to compute $\Pr\{\text{data} \mid \text{model}\}$ and $\Pr\{\text{model} \mid \text{data}\}$, which are the essentials for ecological detection. We provide an ecological scenario for most of the probability distributions, to help make them more concrete. Once again, we encourage you to visit the library and find a mathematics or statistics text that deals with elementary probability theory. Our favorite textbook in introductory probability is by Feller (1968).

We describe four distributions (the binomial, multinomial, Poisson, and negative binomial) in which the random variables are discrete and observations take only integer values. The binomial distribution is commonly used in mark and recapture studies, where a discrete number of individuals are examined. The Poisson is most often used when dealing with counts of the number of plants or animals per unit time or space, or in the analysis of the number of individuals captured. When the data indicate more variability than is consistent with the Poisson distribution, the negative binomial distribution is more appropriate.

We describe four cases in which the random variable is continuous. The first is the normal or Gaussian distribution, which is the commonly used "bell-shaped curve." It has two parameters: the mean and the standard deviation. The normal distribution is commonly used because of a theorem of probability called "the central limit theorem" (Feller 1968), which asserts that, in general (and there are some ecologically important exceptions), when the sum of a large number of random variables is properly scaled (we shall describe this below), the result is approximately normally distributed. This means, for example, that binomial processes with a large number of trials can be approximated by a normally distributed random variable. The normal distribution is symmetric about the mean, which poses many problems in ecology, because this assigns positive probability to values of the random variable that are less than 0, but often the random variable itself (such as length) will have to be greater than 0.

One solution to this problem is to use the log-normal distribution, in which we replace the assumption that the random variable $Z$ has a normal distribution with the assumption that $\log(Z)$—where log denotes the natural logarithm—has a normal distribution. This distribution has an asymmetric shape with a long tail and the property that values of the associated random variable cannot be less than zero. The chi-square distribution is also based on the normal distribution and arises in the study of the distribution of differences between predictions and data.

CHAPTER THREE

TABLE 3.1.  Common probability distributions classified according to the nature of the trials and observations.

|  | Observations | |
|---|---|---|
|  | Discrete | Continuous |
| Trials | | |
| Discrete | Binomial | Normal |
|  |  | Log-normal |
|  |  | Gamma |
| Continuous | Poisson | — |
|  | Negative binomial | |

Finally, we introduce the gamma probability distribution, which is a very flexible continuous distribution that can be used for describing a wide variety of data. It is also an essential component for some of the Bayesian analyses we conduct.

In summary, experiments can involve either discrete or continuous conditions, and the data can be either discrete or continuous (Table 3.1). An overview of these distributions is given in Table 3.2.

### The Binomial Distribution

Perhaps the simplest of probability distribution is the binomial distribution with parameters $N$ and $p$; which we denote by $\mathbf{B}(N,p)$. It arises, for example, in a situation in which an experiment with only two outcomes is repeated $N$ times, and the random variable $Z$ measures the number of times a specified outcome occurs. If $p$ is the chance that the specified outcome occurs in an experiment, then the random variable $Z$ takes integer values ranging from 0 to $N$ according to the rule

$$\Pr\{Z = k\} = p(k,N) = \binom{N}{k} p^k (1 - p)^{N-k}. \tag{3.44}$$

TABLE 3.2. Density, mean, and coefficient of variation of the distributions commonly used by the ecological detective.

| Distribution | Density | Mean | CV |
|---|---|---|---|
| Binomial | $\Pr\{Z = k\} = \binom{N}{k} p^k (1 - p)^{N-k}$ | $Np$ | $\sqrt{\dfrac{1 - p}{Np}}$ |
| Poisson | $\Pr\{Z = k\} = \dfrac{e^{-rt}(rt)^k}{k!}$ | $rt$ | $\sqrt{\dfrac{1}{rt}}$ |
| Negative binomial | $\Pr\{Z = s\} = \dfrac{\Gamma(k + s)}{\Gamma(k)s!} \left(\dfrac{k}{k + m}\right)^k \left(\dfrac{m}{k + m}\right)^s$ | $m$ | $\sqrt{\dfrac{1}{m} + \dfrac{1}{k}}$ |
| Normal | $f(z) = \dfrac{1}{\sqrt{2\pi\sigma^2}} \exp\left(-\dfrac{(z - m)^2}{2\sigma^2}\right)$ | $m$ | $\dfrac{\sigma}{m}$ |
| Gamma | $f(z) = \dfrac{a^n}{\Gamma(n)} e^{-az} z^{n-1}$ | $\dfrac{n}{a}$ | $\dfrac{1}{\sqrt{n}}$ |

In this equation

$$\binom{N}{k} = \frac{N!}{k!(N-k)!}.$$

You should know the following facts about the binomial distribution (Feller 1968). The mean and variance are

$$E\{Z\} = \sum_{k=0}^{N} k \Pr\{Z = k\}$$

$$= Np \quad \text{and} \quad \text{VAR}\{Z\} = Np(1-p). \tag{3.45}$$

The coefficient of variation is

$$\text{CV}\{Z\} = \sqrt{\frac{1-p}{Np}}. \tag{3.46}$$

When $p$ is fixed, the coefficient of variation decreases as $N$ increases. This means that the relative variability shown by $Z$ decreases with the number of experiments conducted.

The values of the binomial probability distribution can be computed by an iterative procedure. First, note that

$$p(0,N) = (1-p)^{N}. \tag{3.47}$$

Then note that $p(k,N)$ and $p(k-1,N)$ can be related as follows:

$$p(k,N) = \binom{N}{k} p^{k}(1-p)^{N-k}$$

$$= \frac{N!}{k!(N-k)!} p^{k}(1-p)^{N-k}$$

$$= \frac{N![N-(k-1)]}{k(k-1)![N-(k-1)]!} p(p^{k-1})(1-p)^{N-(k-1)}(1-p)^{-1}$$

$$= \left[\frac{N-k+1}{k}\right]\left[\frac{p}{1-p}\right] p(k-1,N). \tag{3.48}$$

Equations 3.47 and 3.48 can be implemented in the following manner:

---

Pseudocode 3.1

Step 1.  Specify $p$ and $N$.

Step 2.  Find $p(0,N)$ from Equation 3.47.

Step 3.  For $k = 1$ to $N$, find $p(k,N)$ from Equation 3.48 and print out results in a form that you like.

---

*An Ecological Scenario: Sampling for Pests.*  Suppose that we are sampling fruit for infestations by a pest and know that the chance that a fruit is infested is $p$. If $N$ fruit are sampled, the probability that $k$ of them are infested is given by the binomial distribution. You should use a program based on this pseudocode to predict the distribution of infested fruit if we sample 10 fruit, and $p$ is 0.1, 0.2, or 0.3.

In most situations, we would not know $p$, but need to determine it by sampling fruit. How many fruit should be sampled? How do we estimate $p$ from this sample? What confidence can we associate with this estimate? This becomes a problem in ecological detection that we discuss later.

### The Multinomial Distribution

The multinomial distribution is the extension of the binomial distribution to a case with more than two possible outcomes of the experiment. For example, suppose that the fruit just described could be infested by more than one kind of pest, but there is only one species of pest per fruit. Then the data would be the number of uninfested fruit, the number of fruit infested by pest type 1, the number infested by pest type 2, etc.

Suppose that there are $M$ possible outcomes; we then have a vector of random variables $Z_i$, where $Z_i$ is the number of times the $i^{th}$ kind of outcome occurred. Instead of Equation 3.44 we now consider

67

$$\Pr\{Z_1 = k_1, Z_2 = k_2, \ldots, Z_M$$
$$= k_M \quad \text{in} \quad N \text{ experiments}\}$$
$$= p(k_1, k_2, \ldots, k_M, N), \tag{3.49}$$

which is given by

$$p(k_1, k_2, \ldots k_M, N)$$
$$= \frac{N!}{k_1! k_2! \ldots k_M!} p_1^{k_1} p_2^{k_2} \ldots p_m^{k_m}. \tag{3.50}$$

We encourage you to develop a pseudocode for the multinomial distribution.

### The Poisson Distribution

The binomial distribution is one for which the random variable takes discrete values in discrete experiments or trials. In the same way, the Poisson distribution (or Poisson process, to indicate that something is happening over time) is one for which the random variable takes discrete values during continuous sampling (usually area or time; we use time for definiteness). The Poisson distribution can be derived as the limit of a binomial distribution when $N \to \infty$ and $p \to 0$ in such a way that $Np$ is constant (Feller 1968).

If $Z(t)$ has a Poisson distribution, then

$$\Pr\{Z(t) = k\} = \frac{e^{-rt} (rt)^k}{k!}. \tag{3.51}$$

Here $r$ is called the "rate parameter" of the Poisson distribution. You should know the following facts about the Poisson distribution (Feller 1968).

The mean and variance are

$$E\{Z(t)\} = rt \tag{3.52}$$

and

$$\text{VAR}\{Z\} = rt, \tag{3.53}$$

so that the coefficient of variation is

$$CV\{Z\} = \sqrt{\frac{1}{rt}}. \tag{3.54}$$

Thus, for $r$ fixed, the coefficient of variation decreases as $t$ increases.

The Poisson distribution can be derived from assumptions about what happens in a very small (infinitesimal) amount of time (Feller 1968). Suppose that $\Delta t$ is a very short time interval. We assume that either nothing happens in this time interval or one event happens, and that the probabilities are

$$\Pr\{\text{no event in } \Delta t\} = e^{-r\Delta t},$$
$$\Pr\{\text{ exactly one event in } \Delta t\} = 1 - e^{-r\Delta t}. \tag{3.55}$$

In probability textbooks one usually finds this written as $\Pr\{\text{more than one event in } \Delta t\} = o(\Delta t)$, where $o(\Delta t)$ is the notation that we introduced earlier denoting terms that are high powers of $\Delta t$. Since $e^x = 1 + x + x^2/2 + \ldots$,

$$\Pr\{\text{no events in } \Delta t\} = 1 - r\Delta t + o(\Delta t),$$
$$\Pr\{\text{one event in } \Delta t\} = r\Delta t + o(\Delta t). \tag{3.56}$$

We strongly recommend using Equation 3.55 whenever numerical computation is done, because Equation 3.56 is only an approximation, whereas Equation 3.55 is fundamentally true. For example, regardless of the value of $\Delta t$, Equation 3.56 can lead to probabilities that are bigger than 1 or less than 0 if $r$ is big enough; this does not happen with Equation 3.55.

The mean and variance of the Poisson process are equal. Also, note from Equation 3.55 that the chance of an event in the next bit of time depends only on the time interval and not on any history or current state of the system. We saw this previously with the discussion of random search. Thus, there is a tendency to think of the Poisson distribution as representing "randomness." Since the mean and variance are equal, the tradition evolved in ecology to consider the ratio of the variance of the data to the mean of the data. If this is about 1, then the data are considered to be random, and if the ratio is considerably bigger than 1, then the data are considered to be clumped. Such reasoning only works for special kinds of data, because for this to make sense at all, the data must be dimensionless so that the variance-to-mean ratio has no units.

As with the binomial distribution, it is empowering to be able to compute the terms of the Poisson distribution yourself. This can be done by an iterative procedure. Once again, we begin by setting $p(0,t) = e^{-rt}$. Successive terms are then computed by recognizing that

$$p(k,t) = \frac{e^{-rt}(rt)^k}{k!} = \frac{rt}{k}\frac{e^{-rt}(rt)^{k-1}}{(k-1)!}$$
$$= \left(\frac{rt}{k}\right) p(k-1,t). \tag{3.57}$$

Before we describe the pseudocode, note the following. Unlike the binomial distribution (which has exactly $N$ terms), the Poisson distribution has no limit on the number of terms. Thus, when computing it, you must introduce a cutoff (close to 1), so that when the sum of terms exceeds that cutoff, the computation stops. A pseudocode for this computation is:

---

Pseudocode 3.2
1. Specify $r$, $t$, and the cutoff.
2. Set $p(0,t) = e^{-rt}$. Set sum $= p(0,t)$.
3. Cycle over values of $k \geq 1$ and find $p(k,t)$ from Equation 3.57. Replace sum by sum $+ p(k,t)$.
   If the sum is less than the cutoff, return to
   step 2; otherwise go to step 4.
4. Print out results as you desire.

---

### The Normal or Gaussian Distribution

The two distributions considered thus far involve a random variable $Z$ that takes discrete values. The usual example of a random variable taking continuous values is the normal or Gaussian random variable. We will use the notation $\mathcal{N}(m,\sigma^2)$ to denote a random variable $X$ that is normally distributed with mean $m$ and variance $\sigma^2$. We use the symbol $X$, rather than $Z$, to remind you that these are names of random variables. As long as you remember that they have specific meanings and biological interpretations, there will be no problem.

We need the following facts about the normal distribution. The distribution function $F(x)$ is

$$F(x) = \Pr\{X \leq x\}$$

$$= \frac{1}{\sqrt{2\pi\sigma^2}} \int_{-\infty}^{x} \exp\left(\frac{-(s-m)^2}{2\sigma^2}\right) ds. \tag{3.58}$$

In this expression, the integration variable $s$ takes all values between $s = -\infty$ and $s = x$. Since it must be true that $\Pr\{-\infty < X < \infty\} = 1$,

$$\frac{1}{\sqrt{2\pi\sigma^2}} \int_{-\infty}^{\infty} \exp\left(\frac{-(s-m)^2}{2\sigma^2}\right) ds = 1, \tag{3.59}$$

which means that

$$\int_{-\infty}^{\infty} \exp\left(\frac{-(s-m)^2}{2\sigma^2}\right) ds = \sqrt{2\pi\sigma^2}. \tag{3.60}$$

This is a handy trick for evaluating complicated integrals that are associated with probability functions, and we will use it later.

The normal density function $f(x)$ is

71

$$f(x) = \frac{1}{\sqrt{2\pi\sigma^2}} \exp\left(\frac{-(x-m)^2}{2\sigma^2}\right).$$
$$(3.61)$$

The function $f(x)$ is the familiar "bell-shaped curve." Plot it, if it is not completely familiar; vary $m$ and $\sigma$ to see how they affect the shape.

If $X$ is $\mathcal{N}(m,\sigma^2)$, then the transformed variable $Y = (X - m)/\sigma$ is normally distributed $\mathcal{N}(0,1)$. The distribution function of $Y$ is given the symbol $P_N(y)$:

$$P_N(y) = \frac{1}{\sqrt{2\pi}} \int_{-\infty}^{y} \exp\left(-\frac{s^2}{2}\right) ds,$$
$$(3.62)$$

and once again the integration variable ranges from $s = -\infty$ to $s = y$. This function is especially useful. Note that $P_N(0) = 1/2$ and that if $y < 0$, then $P_N(y) = 1 - P_N(|y|)$.

To find $P_N(y)$ one can compute the value of the integral numerically, but a number of excellent algebraic approximations exist (Abramowitz and Stegun 1965, 932), and we recommend their use. If $y \geq 0$, the following approximation is accurate to $10^{-5}$:

$$P_N(y) = 1 - \frac{1}{\sqrt{2\pi}} \exp\left(-\frac{y^2}{2}\right)[a_1 t + a_2 t^2 + a_3 t^3],$$
$$(3.63)$$

where $t = 1/(1 + py)$, and the constants are $p = 0.332\ 67$, $a_1 = 0.436\ 183\ 6$, $a_2 = -0.120\ 167\ 6$, and $a_3 = 0.937\ 298\ 0$.

It often happens that we want to invert the normal distribution function. That is, we wish to find a value $y_p$ such that $P_N(y_p) = p$, where the value of $p$ is specified. There exist nice algebraic methods for this inversion as well (Abramowitz and Stegun 1965, 933). If $0.5 \leq p \leq 1$, then the following approximation is accurate to $4.5 \times 10^{-4}$:

$$y_p = t - \frac{c_0 + c_1 t + c_2 t^2}{1 + d_1 t + d_2 t^2 + d_3 t^3},$$
$$(3.64)$$

where $t = \sqrt{\log\ (1/p^2)}$, $c_0 = 2.515\ 517$, $c_1 = 0.802\ 853$, $c_2 = 0.010\ 328$, $d_1 = 1.432\ 788$, $d_2 = 0.189\ 269$, and $d_3$

$= 0.001\ 308$. If $p < 0.5$, then we find the value of $y_{1-p}$ according to the same formula and then set $y_p = -y_{1-p}$.

As we mentioned earlier, according to the central limit theorem (CLT), appropriately normalized sums of random variables have a distribution function that approaches the normal distribution. Suppose $\{Z_k\}$ is a sequence of independent random variables with $m_k = E\{Z_k\}$ and $\sigma_k^2 = \text{VAR}\{Z_k\}$, and set

$$S_n = \sum_{k=X}^{n} Z_k,$$

$$m_n = \sum_{k=1}^{n} m_k,$$

$$s_n^2 = \sum_{k=1}^{n} \sigma_k^2. \tag{3.65}$$

According to the CLT, the variable $Z = (S_n - m_n)/s_n$ is approximately normally distributed with mean 0 and variance 1. If the $Z_k$ have the same distribution with common mean $m$ and variance $\sigma^2$, then the $\mathcal{N}(0,1)$ random variable is $(S_n - nm)/\sigma\sqrt{n}$. We shall use the central limit theorem in the next section to motivate the log-normal distribution, and in the next chapter for the determination of the observation effort when monitoring the incidental catch of seabirds in a fishery.

## The Log-Normal Distribution

To understand the log-normal distribution, imagine a population of initial size $N_0$ during a nonbreeding season. We expect the number of individuals alive at some later day $t$, $N_t$, to be the product of $N_0$ and the daily survival probabilities $\{s_i\}$, where $s_i$ is the probability that an individual survives from day $i$ to day $i + 1$. Thus

$$N_t = N_0 s_0 s_1 \cdots s_{t-2} s_{t-1}. \tag{3.66}$$

Taking logarithms of both sides gives

$$\log(N_t) = \log(N_0) + \log(s_0) + \log(s_1)$$
$$+ \cdots + \log(s_{t-1}). \tag{3.67}$$

If the daily survival probabilities are random variables, then using the central limit theorem, we assume that an appropriate normally distributed random variable $Y$ can be constructed from the sum $\log(s_0) + \log(s_1) + \cdots + \log(s_{t-1})$. We then say that $Z = e^Y$ has a log-normal distribution, and we can rewrite Equation 3.66 as

$$N_t = N_0 e^Y = N_0 Z \tag{3.68}$$

One advantage of the log-normal distribution is that a normal random variable takes values between $-\infty$ and $\infty$, but many ecological variables are typically positive. The log-normal random variable takes only positive values. In addition, the log-normal distribution has a long tail, which is common to ecological data.

We will now explore some properties of the log-normally distributed random variable $Z = e^Y$, where we assume that $Y$ is $\mathcal{N}(0, \sigma^2)$. We begin with the distribution function

$$F(z) = \Pr\{Z \le z\} = \Pr\{e^Y \le z\} = \Pr\{Y \le \log(z)\}. \tag{3.69}$$

Since we know that $Y$ is normally distributed with mean 0 and variance $\sigma^2$,

$$F(z) = \frac{1}{\sqrt{2\pi\sigma^2}} \int_{-\infty}^{\log(z)} \exp\left(-\frac{s^2}{2\sigma^2}\right) ds. \tag{3.70}$$

The density function is found by taking the derivative of $F(z)$, and using the chain rule when evaluating the derivative of the integral,

$$f(z) = F'(z) = \frac{1}{\sqrt{2\pi\sigma^2}} \exp\left(-\frac{(\log z)^2}{2\sigma^2}\right) \frac{1}{z}. \tag{3.71}$$

Thus, although $Y$ has a normal density function, the density of $Z$ is skewed, and given by Equation 3.71.

Finally, let us evaluate the mean of the random variable $Z$. Before doing the calculation, we can try to develop some intuition. The mean of the random variable $Y$ is 0, and $Y$ takes positive and negative values. However, $Z = e^Y$ will only take positive values, so that we expect the mean of $Z$ to be larger than 0. We shall now demonstrate this. We start with

$$E\{Z\} = \int_0^\infty zf(z) \ dz = \frac{1}{\sqrt{2\pi\sigma^2}} \int_{-\infty}^\infty e^y \exp\left(-\frac{y^2}{2\sigma^2}\right) dy, \tag{3.72}$$

which is justified by noting that, as $z$ varies from 0 to $\infty$ with density $f(z)$ given by Equation 3.71, $y$ varies from $-\infty$ to $\infty$ with the standard normal density. Bringing the two exponential terms together gives

$$E\{Z\} = \frac{1}{\sqrt{2\pi\sigma^2}} \int_{-\infty}^\infty \exp\left(-\frac{y^2}{2\sigma^2} + y\right) dy. \tag{3.73}$$

We now complete the square in the exponent according to

$$\frac{y^2}{2\sigma^2} - y = \frac{1}{2\sigma^2}[\, y^2 - 2\sigma^2 y]$$

$$= \frac{1}{2\sigma^2}[\, (y - \sigma^2)^2 - \sigma^4\,], \tag{3.74}$$

so that the expected value of $Z$ becomes

$$E\{Z\} = \frac{1}{\sqrt{2\pi\sigma^2}} \int_{-\infty}^\infty \exp\left(-\frac{1}{2\sigma^2}[(y-\sigma^2)^2 - \sigma^4]\right) dy$$

$$= \exp\left(\frac{\sigma^2}{2}\right) \frac{1}{\sqrt{2\pi\sigma^2}} \int_{-\infty}^\infty$$

$$\exp\left(-\frac{1}{2\sigma^2}(y-\sigma^2)^2\right) dy. \tag{3.75}$$

The integrand in Equation 3.75 is a normal density with mean and variance $\sigma^2$, but the range of integration is over all values of $y$; hence the integral must be equal to 1, and we obtain

$$E\{Z\} = \exp\left(\frac{\sigma^2}{2}\right). \tag{3.76}$$

We thus see that the mean of $Z$ is indeed greater than 0, and that the random variable $Z \exp(-\sigma^2/2)$ will have a mean equal to 1. We will use the log-normal distribution extensively in the case studies of fisheries management.

### The Chi-Square Distribution

Another random variable connected to the normal distribution arises as follows. Suppose that the response $Z$ to a control variable $X$ is

$$Z = X + Y, \tag{3.77}$$

where $Y$ is normally distributed with mean 0 and variance 1. The squared deviation between the prediction and the independent variable is then

$$(Z - X)^2 = Y^2, \tag{3.78}$$

and is called the chi-square random variable. If we had $n$ independent variables $\{X_i\}$ and responses $\{Z_i\}$, then the total squared deviation would be

$$\sum_{i=1}^{n} (Z_i - X_i)^2 = \sum_{i=1}^{n} Y_i^2, \tag{3.79}$$

which is called the chi-square random variable with $n$ degrees of freedom and is given the symbol $\chi_n^2$.

### The Gamma Distribution

The gamma distribution also takes non-negative values, can have a long tail, and is very useful in Bayesian analysis.

A random variable $Z$ follows a gamma density with parameters $a$ and $n$ if the probability density function is

$$f(z) = \frac{a^n}{\Gamma(n)} e^{-az} z^{n-1}. \qquad (3.80)$$

In this equation, $\Gamma(n)$ is read "gamma of $n$" and is described in Box 3.3.

If you don't worry about these things, just think of $a^n/\Gamma(n)$ as a normalization constant to ensure that $f(z)$ defined in Equation 3.80 is a true probability density (i.e., its integral is 1); the gamma function plays the same role that $\binom{N}{k}$ plays in the binomial distribution. That is, since $\Pr\{0 \leq Z < \infty\}$,

$$\int_0^\infty \frac{a^n}{\Gamma(n)} e^{-az} z^{n-1} \, dz = 1. \qquad (3.81)$$

Since $a^n/\Gamma(n)$ is a constant, it can be brought out of the integral sign:

$$\frac{a^n}{\Gamma(n)} \int_0^\infty e^{-az} z^{n-1} \, dz = 1$$

$$\text{or} \quad \int_0^\infty e^{-az} z^{n-1} \, dz = \frac{\Gamma(n)}{a^n}. \qquad (3.82)$$

We now consider some properties of the gamma density, Equation 3.80. To begin, note that if $n = 1$, since $\Gamma(1) = 0! = 1$, $f(z) = e^{-az}$, which is the exponential density.

When $n < 1$, as $z \to 0$, $z^{n-1} \to \infty$, so that $f(z) \to \infty$. When $n > 1$, $z^{n-1}$ will approach 0 as $z \to 0$, so that $f(0) = 0$ and the gamma density has a peak (Figure 3.5) because $e^{-az} \to 0$ as $z$ increases. Thus, the single parameter $n$ controls the wide-ranging shape of this density.

---

BOX 3.3

AN ASIDE ON THE GAMMA FUNCTION

Most readers will feel comfortable with the more common special functions such as $\log(x)$, $e^x$, or $\sin(x)$ and $\cos(x)$. These relatively simple transcendental functions (i) are encountered frequently, (ii) often have simple physical interpretations, (iii) are well tabulated, and (iv) have simple power series and limiting behaviors, such as $x^n/e^x \to 0$ as $x \to \infty$ for any $n$, or $\lim_{x \to 0} (\sin x)/x = 1$.

The gamma function shares many of the same qualities. A good source book is by Abramowitz and Stegun (1965). The gamma function $\Gamma(n)$ arises in classical applied mathematics, and is defined by the integral

$$\Gamma(n) = \int_0^\infty e^{-t} t^{n-1} \, dt.$$

Integrating by parts gives

$$\Gamma(n+1) = \int_0^\infty e^{-t} t^n \, dt = -e^{-t} t^n \big|_0^\infty$$

$$+ \int_0^\infty e^{-t} n t^{n-1} \, dt = n\Gamma(n),$$

so that we conclude that

$$\Gamma(n + 1) = n\Gamma(n).$$

This recurrence formula is similar to the one for factorials in which $n! = n(n - 1)!$ For integer values of $n$, $\Gamma(n + 1) = n!$ The general recurrence holds for all values of $n$, however, not just integer ones.

BOX 3.3 CONT.

Abramowitz and Stegun (1965, 256) show that $\Gamma(n)$ can be calculated from the formula

$$\Gamma(n) = \frac{1}{\displaystyle\sum_{k=1}^{\infty} c_k n^k}.$$

The best way to use this formula is to write $n = n_I + n_F$, where $n_I$ is an integer and $n_F$ is a fraction with $0 < n_F < 1$. First compute $\Gamma(n_F)$ from the power series and then use the recurrence relationship. For example, $\Gamma(3.7) = 2.7\Gamma(2.7) = (2.7)(1.7)\Gamma(1.7) = (2.7)(1.7)(0.7)\Gamma(0.7)$. The first nineteen of the $c_k$ are (Abramowitz and Stegun 1965, 256):

| $k$ | $c_k$ | | | | | |
|---|---|---|---|---|---|---|
| 1 | 1.000 | 000 | 000 | 000 | 000 | 0 |
| 2 | 0.577 | 215 | 664 | 901 | 532 | 9 |
| 3 | $-0.655$ | 878 | 071 | 520 | 253 | 8 |
| 4 | $-0.042$ | 002 | 635 | 034 | 095 | 2 |
| 5 | 0.166 | 538 | 611 | 382 | 291 | 5 |
| 6 | $-0.042$ | 197 | 734 | 555 | 544 | 3 |
| 7 | $-0.009$ | 621 | 971 | 527 | 877 | 0 |
| 8 | 0.007 | 218 | 943 | 246 | 663 | 0 |
| 9 | $-0.001$ | 165 | 167 | 591 | 859 | 1 |
| 10 | $-0.000$ | 215 | 241 | 674 | 114 | 9 |
| 11 | 0.000 | 128 | 050 | 282 | 388 | 2 |
| 12 | $-0.000$ | 020 | 134 | 854 | 780 | 7 |
| 13 | $-0.000$ | 001 | 250 | 493 | 482 | 1 |
| 14 | 0.000 | 001 | 133 | 027 | 232 | 0 |
| 15 | $-0.000$ | 000 | 205 | 633 | 841 | 7 |
| 16 | 0.000 | 000 | 006 | 116 | 095 | 0 |
| 17 | 0.000 | 000 | 005 | 002 | 007 | 5 |
| 18 | $-0.000$ | 000 | 001 | 181 | 274 | 6 |
| 19 | 0.000 | 000 | 000 | 104 | 342 | 7 |

(a)

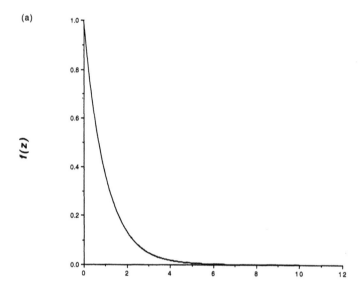

(b)

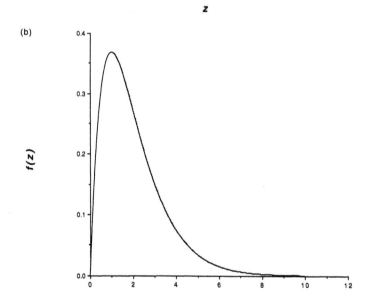

(c)

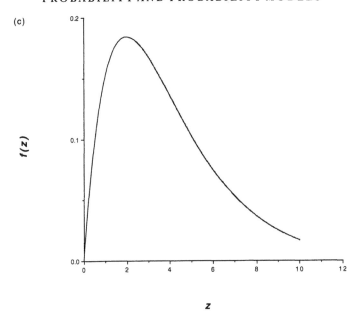

FIGURE 3.5. The gamma probability density $f(z)$ for three values of the parameters $n$ and $a$. Note the range of shapes that is possible for this density. (a) $n = 1$, $a = 1$; (b) $n = 2$, $a = 1$; (c) $n = 2$, $a = 0.5$.

The expected value of $Z$ is

$$E\{Z\} = \int_0^\infty zf(z)\ dz = \int_0^\infty \frac{a^n}{\Gamma(n)}\ e^{-az}z^n\ dz.$$
(3.83)

Using a modification of Equation 3.82 gives

$$E\{Z\} = \frac{\Gamma(n+1)}{a^{n+1}} \frac{a^n}{\Gamma(n)} = \frac{n}{a}.$$
(3.84)

The mean of the gamma density is the ratio of the parameters. The most likely value (i.e., the "mode") of the gamma density is found by setting the derivative of $f(z)$ equal to 0 and solving for $z^* = (n - 1)/a$, so that the most likely value of the gamma density occurs at a value smaller than the mean, and therefore the density has a long tail.

81

We encourage you to find the second moment using the same method, and then show that the coefficient of variation is

$$CV\{Z\} = \frac{1}{\sqrt{n}}, \tag{3.85}$$

so that the single parameter $n$ also controls the coefficient of variation.

*Ecological Scenario: The Return to Random Search.* We now return to the discussion of random search by predators (Box 3.2). Recall that we concluded there that

$$Q(t) = \Pr\{\text{no food is found between 0 and } t\} = e^{-ct}, \tag{3.86}$$

and that this function has the memoryless property that unsuccessful search up to time $t$ provides no information about the chance of success after that time.

Previously, we assumed that $c$ was a fixed constant. Let us now suppose, however, that $c$ has a frequency distribution. For example, the search rate might vary across seasons, across spatial locations as the predator searches, or across individual prey items. In that case, Equation 3.86 is reinterpreted as the conditional probability of not finding food, given the value of $c$. Assume that $c$ has a gamma density. Then the joint probability of not finding food and the value of $c$ is

$\Pr\{\text{no food is found between 0 and } t \text{ and the}$
search parameter takes the value $c\}$

$$= e^{-ct} \frac{a^n}{\Gamma(n)} e^{-ac} c^{n-1}. \tag{3.87}$$

Consequently, the probability of not finding food is

$\Pr\{\text{no food is found between 0 and } t\}$

$$= \int_0^\infty \Pr \left\{ \begin{array}{l} \text{no food is found between 0 and } t \\ \text{and the search parameter takes} \\ \text{the value } c \end{array} \right\} dc$$

$$= \int_0^\infty \frac{a^n}{\Gamma(n)} e^{-(a+t)c} c^{n-1} \, dc = \left( \frac{a}{a+t} \right)^n.$$

(3.88)

You should verify the integral, once again by using logic similar to that in Equation 3.84. You should also verify, following the same calculation as in Box 3.2, that the distribution in Equation 3.88 does not have the memoryless property. We return to this example once more, when we discuss Bayesian analysis, because if $c$ has a distribution of values, the predator can learn from its failed search and learning changes the frequency distribution. The precise way that this is done requires the methodology introduced in Chapter 9.

### The Negative Binomial Distribution

The negative binomial distribution arises in two ways, and both are relevant to the ecological detective. First, imagine a sequence of independent experiments, each of which has probability $p$ of succeeding. We are interested in the number of experiments needed before $s$ successes occur. In particular, we ask for the probability that the $s^{th}$ success occurs on trial $Z = u + s$, where $u$ is the number of unsuccessful experiments, so that $u = 0, 1, 2, \ldots$. The $s^{th}$ success can happen on trial $u + s$ only if there are $s - 1$ successes in the first $u + s - 1$ experiments and a success on the $(u + s)^{th}$ experiment. The probability of the latter event is $p$ and the probability of the former is given by the binomial distribution

$$\binom{u + s - 1}{s - 1} p^{s-1}(1 - p)^{u+s-1-(s-1)}$$

$$= \binom{u + s - 1}{u} p^{s-1}(1 - p)^u.$$

(3.89)

Multiplying this expression by $p$, we obtain

Pr$\{s^{th}$ success occurs on trial $u + s\}$

83

$$= \begin{pmatrix} u + s - 1 \\ u \end{pmatrix} p^s (1 - p)^u. \quad (3.90)$$

This is the first form of the negative binomial distribution. Here the parameters are $u$ and $p$ with the possible values $u > 0$ and $0 < p < 1$.

The second form of the negative binomial distribution arises when we consider a Poisson process in which the rate parameter has a probability distribution. In that case, we can interpret Equation 3.51 as a conditional probability:

$$\Pr\{Z(t) = s \mid \text{parameter} = r\} = \frac{e^{-rt}(rt)^s}{s!}. \quad (3.91)$$

Now assume that $r$ has a gamma density with parameters $n$ and $a$, so that the expected value of $r$ is $n/a$. The unconditional distribution of $Z(t)$ is found by integrating the product of the conditional distribution Equation 3.91 and the gamma density, since this product is the $\Pr\{Z(t) = s$ and the parameter $= r\}$, over all possible values of $r$:

$$\Pr\{Z(t) = s\} = \int_0^\infty \frac{e^{-rt}(rt)^s}{s!} \frac{a^n}{\Gamma(n)} e^{-ar} r^{n-1} \, dr. \quad (3.92)$$

Taking everything that is constant out of the integral gives

$$\Pr\{Z(t) = s\} = \frac{t^s a^n}{s!\Gamma(n)} \int_0^\infty e^{-(t+a)r} r^{s+n-1} \, dr. \quad (3.93)$$

Computing the integral as before,

$$\int_0^\infty e^{-(t+a)r} r^{s+n-1} \, dr = \frac{\Gamma(n+s)}{(a+t)^{n+s}}, \quad (3.94)$$

so that

$$\Pr\{Z(t) = s\} = \frac{t^s a^n}{s!\Gamma(n)} \frac{\Gamma(n+s)}{(a+t)^{n+s}}$$

$$= \frac{\Gamma(n + s)}{\Gamma(n)} \frac{t^s}{s!} \frac{a^n}{(a + t)^{s+n}}$$

$$= \frac{\Gamma(n + s)}{\Gamma(n)s!} \left(\frac{t}{a + t}\right)^s \left(\frac{a}{a + t}\right)^n. \tag{3.95}$$

If we set $p = a/(a + t)$, then Equation 3.95 can be rewritten as

$$\Pr\{Z(t) = s\} = \binom{n + s - 1}{s} p^n (1 - p)^s, \tag{3.96}$$

which is analogous to Equation 3.90 with $n$ replacing $u$. The difference is that we now allow any value of $n$, whereas in Equation 3.90 the understanding is implicitly that $u$ is at least 1.

The mean of the negative binomial distribution is

$$E\{Z(t)\} = \frac{n(1 - p)}{p} = \frac{n}{a} t = m(t) \tag{3.97}$$

and the variance is

$$\text{VAR}\{Z(t)\} = m(t) + \frac{m(t)^2}{n}. \tag{3.98}$$

Unlike the case of the Poisson distribution, in which the variance and mean are equal, the variance of the negative binomial distribution will always be larger than the mean. Hence, $n$ is often called the "overdispersion" parameter. We can see this more clearly by considering the coefficients of variation. For the Poisson distribution,

$$\text{CV}_{\text{Poisson}}\{Z(t)\} = \frac{1}{\sqrt{rt}}, \tag{3.99}$$

whereas for the negative binomial distribution,

$$\text{CV}_{\text{NB}}\{Z(t)\} = \sqrt{\frac{1}{m(t)} + \frac{1}{n}}. \tag{3.100}$$

From Equation 3.97 we see that as $t \to \infty$, $m(t) \to \infty$. Although the CV of the Poisson distribution goes to 0 as $t \to \infty$, the CV of the negative binomial distribution approaches a constant. Note that as $n$ increases, $CV_{NB}$ approaches $CV_{Poisson}$. This can be shown more precisely: that as $n \to \infty$, the negative binomial distribution becomes more and more Poisson-like.

A form of the negative binomial distribution commonly encountered in ecological texts (e.g., Southwood 1966), and one that we find handy to use, is

$$\Pr\{Z(t) = s\} = \frac{\Gamma(k + s)}{\Gamma(k)\, s!} \left( 1 + \frac{m}{k} \right)^{-k} \left( \frac{m}{m + k} \right)^{s}, \quad (3.101)$$

where $k$ and $m$ are parameters. Using Equation 3.95, setting $m(t) = (n/a)\, t$, and doing some algebra shows that

$$\Pr\{Z(t) = s\}$$
$$= \frac{\Gamma(n + s)}{\Gamma(n)\, s!} \left( \frac{m(t)}{n + m(t)} \right)^{s} \left( \frac{n}{n + m(t)} \right)^{n}. \quad (3.102)$$

Comparing Equations 3.101 and 3.102, we see that $m(t)$ and $m$ have exactly the same interpretation as the mean, and that $k$ and $n$ have exactly the same interpretation as the overdispersion parameter.

We can find the terms of the negative binomial distribution using an iterative procedure similar to the one used for the binomial and Poisson distributions. For purposes of commonality with most ecological texts, we adopt Equation 3.102, rewriting it with $Z(t) = Z$, $m(t) = m$, and $n = k$, so that

$$\Pr\{Z = s\} = \frac{\Gamma(k + s)}{\Gamma(k)\, s!} \left( \frac{m}{k + m} \right)^{s} \left( 1 + \frac{m}{k} \right)^{-k}, \quad (3.103)$$

and note that the last term is the same as $[k/(k + m)]^{k}$ so that

$$\Pr\{Z = s\} = p(s,m,k)$$

$$= \frac{\Gamma(k + s)}{\Gamma(k)\,s!} \left( \frac{m}{k + m} \right)^{s} \left( \frac{k}{k + m} \right)^{k}. \qquad (3.104)$$

From this equation, we see that $p(0,m,k) = [k/k + m)]^{k}$ and that additional terms can be computed according to

$$p(s,m,k) = \frac{s + k - 1}{s} \frac{m}{k + m} p(s - 1,m,k)$$

$$\text{for } s = 1, 2, \ldots. \qquad (3.105)$$

The iteration result, Equation 3.105, is derived in the same way as the iteration results for the binomial and Poisson distributions were derived. We encourage you to derive it and write out the pseudocode. We shall now use it.

## THE MONTE CARLO METHOD

In order to confront models with data, we must estimate parameters in the models from the data and then choose one description of nature over another. Because we usually do not know the true mechanisms and processes in the natural world, we never know if the parameters that we estimate are indeed "true" or if the model that is picked is "correct." One way to increase our confidence in the methods we use is to test models and methods on sets of data in which we know exactly what is happening, i.e., where we create the data and thus know the true situation exactly. A useful method for generating such data is called the Monte Carlo method or the method of stochastic simulation (Ripley 1987).

The Monte Carlo method uses random-number generators for the construction of data. Virtually all microcomputer languages have built-in random-number generators, and these are, for almost all of our purposes, sufficient. The usual problem with such generators is that they are only quasi-random and have a periodic cycling in the generation of the numbers. These days, however, the periods are of the order of $2^{30}$, so that the difficulties are minor. The random number generators usually provide a value $U$ that is uniformly distributed between 0 and 1. Thus, the distribution function for $U$ is

$$F(u) = \begin{cases} u & \text{if } 0 \le u \le 1, \\ 0 & \text{otherwise,} \end{cases} \qquad (3.106)$$

and the density is $f(u) = F'(u) = 1$.

To construct a random variable $Z$ that is uniformly distributed on the interval $[A,B]$, we pick $U$ and set

$$Z = A + (B - A)U. \qquad (3.107)$$

Since the smallest value that $U$ takes is 0, the smallest value that $Z$ takes is $A$; similarly, the largest value of $Z = B$, corresponding to $U = 1$.

Typically, the command $U = $ RND in a computer program will generate a uniformly distributed random variable (but check the manual for your software). We now describe methods for generating random variables with other distributions.

### *Binomial, Poisson, or Negative Binomial Random Variables*

These three distributions have the common feature that the random variable $Z$ takes integer values. We shall illustrate the method for generating individual random variables from a specific distribution using the binomial distribution, and leave the Poisson and negative binomial distributions to you.

For the binomial distribution, the probability $p(k,N)$ of obtaining exactly $k$ successes in $N$ experiments is given by Equation 3.44. If $p(k,N)$ is summed from $k = 0$ to $k = N$, the sum is 1. The value of $k$ associated with a particular value of $U = u$, called $k_u$, is chosen so that

$$\sum_{k=0}^{k_u} p(k,N) \leq U$$

and

$$\sum_{k=0}^{k_u+1} p(k,N) > U. \tag{3.108}$$

A pseudocode that implements this idea is:

---

Pseudocode 3.3

1. Specify parameters $N$ and $p$. Choose a uniformly distributed random number $U$. Set $k = 0$ and SUM $= 0$.
2. Compute $p(k,N)$ from Equation 3.44.
3. Replace SUM by SUM $+ p(k,N)$.
4. If SUM $\geq U$, then the current value of $k$ is the number of successes in this single experiment. Otherwise, replace $k$ by $k + 1$ and return to step 2.

---

### Normal Random Variables

To generate normally distributed random variables, we recommend the use of the Box-Mueller scheme (Press et al. 1986, 202). Choose two uniformly distributed random numbers $U_1$ and $U_2$ and set

$$Z_1 = \sqrt{-2 \log(U_1)} \, \cos(2\pi U_2),$$

$$Z_2 = \sqrt{-2 \log(U_1)} \, \sin(2\pi U_2). \tag{3.109}$$

Then $Z_1$ and $Z_2$ are normally distributed random variables with mean 0 and variance 1. To make these variables normally distributed with mean $m$ and variance $\sigma^2$, replace $Z_i$ by $m + \sigma Z_i$. We leave writing a pseudocode to you.

## Gamma Random Variables

Gamma random variables are more difficult to generate. Press et al. (1986, 204 ff.) describe one method ("the rejection method") that interested readers may wish to consult. In general, some form of integration of the probability density is needed.

### An Ecological Scenario: The Simple Population Model with Process and Observation Uncertainty

We return to the model Equation 3.38. In order to generate data with this model, assumptions about $W_t$, $V_t$, and the other parameters are required. For example, we might assume that the process and observation uncertainties are normally distributed with mean 0 and standard deviations $\sigma_W$ and $\sigma_V$, respectively, but that the initial population size $N_0$ is known exactly. As a demonstration of the importance of understanding observation and process uncertainty, and to demonstrate the Monte Carlo technique, we now perform some simple computer experiments based on the following pseudocode. A pseudocode for this model with process and observation uncertainties is:

---

Pseudocode 3.4

1. Specify $s$, $b$, $\sigma_N$, $\sigma_W$, $\sigma_V$, and $N_0$.
2. Begin a loop over 50 time steps.
3. Calculate $N_{t+1}$ and $N_{obs,t}$ from Equation 3.38.
4. Print or graph results as desired.
5. Exit after 50 time steps.

---

We chose $s = 0.8$, $b = 20$, and $N_0 = 50$.

To begin, we can ask how process uncertainty affects the relationship between $N_t$ and $N_{t+1}$. If we allow for process uncertainty ($\sigma_W = 10$), but no observation uncertainty ($\sigma_V = 0$), the observed values are "scattered" about the true value (Figure 3.6) but will be centered on it. A standard

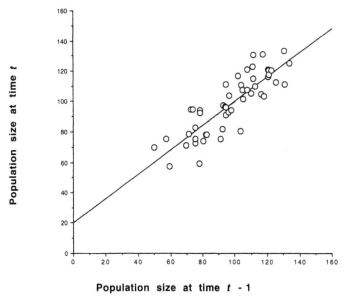

FIGURE 3.6. One Monte Carlo realization of fifty data points drawn with process uncertainty but no observation uncertainty. The solid line represents the true model (deterministic relationship).

linear regression fit to the data gives $y = 20.01 + 0.808x$ with $r^2 = 0.723$. Thus, both the birth rate (the constant in the regression) and the survival (the slope of the regression) are accurately determined.

If we now add observation uncertainty, by setting $\sigma_V = 10$, and use the same sequence of random numbers to generate the data, we obtain an apparent "relationship" (Figure 3.7) that is weaker than in the case without observation uncertainty. In this case, the regression is $y = 32.47 + 0.684x$ with $r^2 = 0.481$. Thus, we overestimate the birth rate, underestimate survival, and explain only about half as much of the variation as before. What happened? By adding variability in observations, it now appears that there are some very small population sizes and some very large ones, even

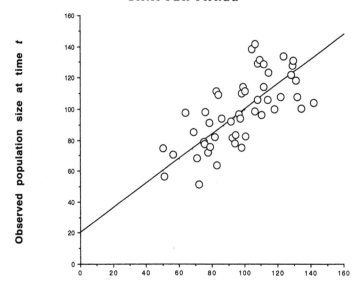

Observed population size at time $t$ - 1

FIGURE 3.7. One Monte Carlo realization of fifty data points drawn with process and observation uncertainty. Once again, the solid line represents the true model (deterministic relationship).

though the true population size has not changed. The net effect is that the population in the next time period, $N_{t+1}$, appears to depend less on $N_t$. This is not due to a weakening of the density dependence. Rather, it is caused by the additional source of uncertainty in the model. The job of the ecological detective is to sort out such differences and then arrive at the best description of nature possible.

### Bootstrap Data Sets

Another use of the Monte Carlo method is to generate "replicate" sets of data from one actual set of data. This is often called a "bootstrap" data set (Efron and Tibshirani 1991, 1993). We do it by resampling the data set with replacement. For example, in the discussion of coefficient of variation, we described a data set of masses of rodents. The original data (in g) were {79,120,85,99,100}. A bootstrap data set is constructed by randomly picking five "new" masses from the original data set, with replacement. Thus, one such bootstrap replicate might be {79,120,85,100,100} and another might be {99,75,99,120,85}. We could use this method to generate a large number of "replicate" data sets. We will use the bootstrap method for both model selection and the evaluation of confidence limits.

# Incidental Catch in Fisheries: Seabirds in the New Zealand Squid Trawl Fishery

## MOTIVATION

It often happens that nontarget species are captured during fishing operations. These takes are called "incidental catch." In some cases, such as the high-seas driftnet fisheries (Mangel 1992), large-scale observer programs are used to monitor incidental catch. Questions arise about how to set the level of observer coverage and how to interpret the data collected during the observer programs. In this chapter, we analyze a particular fishery and compare the conclusions obtained using different models to describe the incidental catch. This example demonstrates the importance of knowing your data, application of the central limit theorem, and how the Monte Carlo technique can be used to make predictions.

## THE ECOLOGICAL SETTING

In the trawl fishery for squid in the waters off New Zealand,

Seabirds, and especially albatrosses, naturally scavenge for dead squid at the sea surface . . . and several species have learnt to recognize trawlers and to specialize in scavenging trawl waste. . . . The best time for seabirds to obtain food from trawlers is when the net is being hauled, or as

94

the waste is being discharged from the factory [ship]. Sea-birds sometimes become entangled in the trawl net itself (cod-end and wings), and sometimes in trawl gear such as the floatline or bellylines. However, of the albatrosses caught in the squid trawl fishery, at least 82% and proba-bly 93% are killed by collision (and entanglement) with the netsonde monitor cable. (Bartle 1991)

The netsonde cable is old equipment; most modern vessels use hull-mounted transducers or tow aquaplanes, and this helps reduce incidental mortality.

The data were collected by eleven observers on four dif-ferent vessels during the 1990 fishing season. The observers worked for the New Zealand Ministry of Agriculture and Fisheries; they collected data over 338 days of fishing and observed 897 of 4349 tows in that season. According to Bar-tle (1991), the general pattern of capture rates and types of animals captured is representative of other years. Observers recorded positions of the vessels, tow numbers, and num-bers and types of birds captured incidentally (snagged in the netsonde monitor cable or entangled in the net and fishing gear). This fishery was almost closed in 1992 because of incidental mortality of Hooker's sea lion; incidental take is not restricted to birds.

Seven species of birds were taken incidentally: the royal albatross *Diomedea epomorphora* (1 animal captured inciden-tally), grey-headed albatross *Diomedea chrysostoma* (3 animals captured incidentally), Buller's albatross *Diomedea bulleri* (3), white-capped albatross *Diomedea cauta steadi* (250), white-chinned petrel *Procellaria aequinoctialis* (2), sooty shearwater *Puffinus griseus* (30), and Prion *Pachyptila* sp. (4). The focus of Bartle's work, and of ours, is the white-capped albatross, which is the only species caught in sufficient numbers to be of considerable concern.

95

## "STATISTICALLY MEANINGFUL DATA"

In observer programs, one of the most important questions involves setting the level of observer coverage to obtain "statistically meaningful data." For logistical and funding reasons, setting observer levels must usually be done in advance of the season. The typical starting point is the assumption (by an appeal to the central limit theorem) that the number of animals killed per operation is normally distributed with mean $\mu$ and variance $\sigma^2$, which are unknown. We determine the necessary observer level (i.e., the number of tows observed) in the current year, given data from the previous year, to meet a specified criterion of accuracy.

Suppose that $N_{\text{last}}$ tows were observed last year and that $c_i$ was the number of animals killed on the $i^{\text{th}}$ tow. We assume that the sample mean,

$$m = \frac{1}{N_{\text{last}}} \sum_{i=1}^{N_{\text{last}}} c_i,$$

(4.1)

provides an estimate of $\mu$. Similarly, the sample variance,

$$s'^2 = \frac{1}{N_{\text{last}} - 1} \sum_{i=1}^{N_{\text{last}}} (c_i - m)^2,$$

(4.2)

is assumed to provide an estimate of $\sigma^2$. The denominator, $N_{\text{last}} - 1$, is used to correct for bias in the estimate of variance (e.g., Kendall and Stuart 1979).

The objective is to determine the number of tows $N$ to be observed this year so that the mean kill can be estimated with a given level of accuracy, assuming that the distribution of by-catch this year is the same as it was last year. This assumption allows us to apply $m$ and $s'^2$ to estimate $\mu$ and $\sigma^2$. One measure of accuracy is the width of the confidence interval of the mean. Because the variance is unknown, confi-

TABLE 4.1. Commonly used values of Student's $t$.

| Confidence level | $t$ value ($t_q$) |
|---|---|
| 90% | 1.645 |
| 95% | 1.960 |
| 99% | 2.576 |
| 99.9% | 3.291 |

dence intervals for the mean are constructed using Student's $t$ distribution (Kendall and Stuart 1979). The statistical theory of sampling shows that confidence intervals for the mean are of the form

$$m - \frac{s'}{\sqrt{N}} t_{q,N-1} \leq \mu \leq m + \frac{s'}{\sqrt{N}} t_{q,N-1},\qquad (4.3)$$

where $t_{q,N-1}$ is Student's $t$ value corresponding to a given confidence level and $N - 1$ degrees of freedom. For the sample sizes we are talking about, we can equate $t_{q,N-1}$ with $t_{q,\infty}$ and call it $t_q$ when looking up values of Student's $t$ (Table 4.1). In general, too, when $N > 30$ or so, the resulting confidence intervals are essentially the same as those that would result if the true variance were used.

The width $W$ of the confidence interval is

$$W = 2 \frac{s'}{\sqrt{N}} t_q. \qquad (4.4)$$

Suppose that we define a "statistically accurate" observer program as one that determines an estimate to within half of the width of the confidence interval (this is often done). We call this the "tolerable error of the mean," and denote it by $d$:

$$\text{Tolerable error of the mean} = d = \frac{s'}{\sqrt{N}} t_q. \qquad (4.5)$$

That is, an acceptable level of observer coverage is one for which the deviation between the true value of the mean and the estimate is never more than $d/2$, with the value of $d$ specified in advance.

Once $d$ is specified, the required level of observer coverage is

$$N = \frac{s'^2}{d^2} t_q^2.$$

(4.6)

This procedure was used, for example, by scientists from the United States, Canada, and Japan in estimating observer level needed in high-seas driftnet fisheries (Mangel 1992). Note that although $s'^2$ comes from the data, the values of $d$ and $t_q$ have to be agreed on in advance. For example, the difference between using a 90% confidence level (for which $t_q = 1.645$) and a 95% level (for which $t_q = 1.960$) will be a difference of $(1.96/1.645)^2 = 1.42$, or 42% in the required levels of observer coverage. We will see that the choice of the value of $s'^2$ to use in the computation is a problem of ecological detection.

## THE DATA

Observer programs typically generate considerable amounts of data, so that the results are usually presented in summarized form. Bartle (1991) does this too, and those data are given in Table 4.2. Note that there is about a four-fold difference in by-catch rates from the different vessels. The average by-catch rate, over all vessels and observation periods, is 0.215 animals/tow and the sample standard deviation is 0.141. Note that this is different from the value (0.263 birds/tow) that we would obtain by dividing the total number of animals killed (236) by the total number of tows observed (897). You should think about why this is so. The data in Table 4.2 represent the first of two models that will be used.

The data in Table 4.2 are summarized over observation periods as long as four months. But the fundamental vari-

TABLE 4.2. Summary of observer data.

| Dates[a] | Tows observed | Number of white-capped albatrosses captured | Capture rate (birds/tow) |
|---|---|---|---|
| 31 Dec–3 Mar | 237 | 72 | 0.304 |
| 7 Jan–12 Feb | 125 | 36 | 0.288 |
| 15 Jan–24 Apr | 240 | 100 | 0.417 |
| 30 Jan–23 Feb | 73 | 8 | 0.110 |
| 28 Feb–24 Mar | 70 | 5 | 0.071 |
| 17 Feb–25 Apr | 152 | 15 | 0.099 |

[a]For the 1989–90 fishing season. These data are from five different vessels. (The 30 Jan–23 Feb and 28 Feb–24 Mar observations were on the same vessel.)

able of interest to us is the number of birds incidentally captured in a single tow. Since the average by-catch rate is less than one animal per tow, there must be instances in which no animals were captured. One way of generating a rate of about 0.25 animals/tow would be to have about three tows with no by-catch for each tow with one animal captured. But there could also be twenty tows with no animals and one tow with five or six animals. The difference between these two cases is important. In the first case, the mean catch rate is much more representative of the data than in the second case. In order to investigate this issue further, we must consider the original, rather than summarized, data. Luckily, Bartle (1991, Table 4) gives the unsummarized data, which we reproduce here in Table 4.3. For these data the average by-catch is 0.279 birds/tow but the standard deviation is 1.25—nearly ten times larger than the standard deviation for the summarized data. Note too that the frequency of no birds being captured on a tow is about 90% and that, of the birds captured, about 65% were captured with a rate greater than or equal to three birds per tow (also noted by Bartle 1991). Thus, the average capture

99

TABLE 4.3. Frequency distribution of by-catch. Data from Bartle (1991). Bartle actually reports the zero category as "371$^{+}$"; we have computed the value here from the total number of tows observed (Table 4.2).

| Number of albatrosses captured | Number of hauls with this level of by-catch |
|:---:|:---:|
| 0 | 807 |
| 1 | 37 |
| 2 | 27 |
| 3 | 8 |
| 4 | 4 |
| 5 | 4 |
| 6 | 1 |
| 7 | 3 |
| 8 | 1 |
| 9 | 0 |
| 10 | 0 |
| 11 | 2 |
| 12 | 1 |
| 13 | 1 |
| 14 | 0 |
| 15 | 0 |
| 16 | 0 |
| 17 | 1 |

rate is unlikely: either no birds are captured or quite a few are. This is our second model. It surely is a more accurate representation and the prediction about the level of observer coverage needed is very, very different. Using the two standard deviations in Equation 4.6—where they enter as squares—shows that there is about an eightyfold difference in the amount of required observer coverage predicted!

## A NEGATIVE BINOMIAL MODEL OF BY-CATCH

A negative binomial model can be used for the case in which the average by-catch is unlikely, i.e., where the by-catch is highly aggregated so that no by-catch is a common occurrence and high levels of by-catch are a rare occurrence. We adopt the "$m,k$" version of the negative binomial distribution for the probability that the by-catch on the $i^{th}$ tow, $C_i$, equals a particular value $c$:

$$\Pr\{C_i = c\} = p(c) = \frac{\Gamma(k + c)}{\Gamma(k)\,c!} \left(\frac{k}{k + m}\right)^k \left(\frac{m}{m + k}\right)^c. \quad (4.7)$$

We can estimate $m$ and $k$ by the nonlinear search techniques described in Chapter 11. We can also estimate the parameters by the method of moments. Recall that the mean and variance of $C_i$ are

$$E\{C_i\} = m,$$

$$\mathrm{VAR}\{C_i\} = m + \frac{m^2}{k}, \quad (4.8)$$

so that the sample mean provides an estimate of $m$, and an estimate of $k$ is obtained by solving Equation 4.8 for $k$ using the sample mean and variance (this is called the moment estimator). For the data in Table 4.3, $k = 0.06$. Recall that when the overdispersion parameter $k \to \infty$, the negative binomial distribution is approximated by the Poisson distribution (for which the mean is a likely observation); here we are in the other extreme and the negative binomial model gives a very good description of the data (Figure 4.1). We encourage you to experiment with the Poisson model and the data.

There is one more step. We should demonstrate that the NB model is also absolutely likely. To do this, we compute the standard chi-squared test. If $N_{\mathrm{tow}}$ is the total number of tows observed, and $n(c)$ tows are observed with incidental catch level $c$, then $n(c)/N_{\mathrm{tow}}$ is the observed frequency of incidental catch $c$ and $p(c)$ is the expected frequency. Thus the chi-squared variate is

(a)

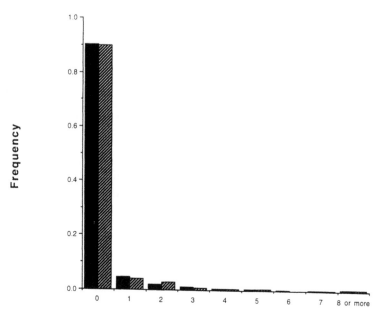

Frequency

Level of incidental catch

(b)

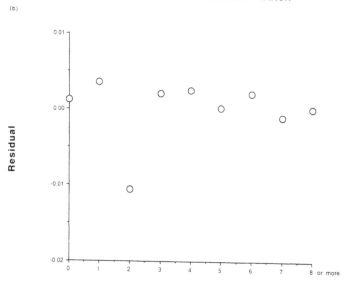

Residual

Level of incidental catch

$$\chi^2 = \sum_{c=0}^{17} \frac{[n(c)/N_{tow} - p(c)]^2}{p(c)}.$$

(4.9)

When computing this quantity, we run into one difficulty. If $p(c) = 0$, which may occur for large values of by-catch, the denominator in Equation 4.9 is 0. To get around this problem, we can either limit the sum in Equation 4.9 or pool the large values of $c$ into a single category. The resulting value of $\chi^2$ corresponds to probabilities of 0.2–0.4, depending upon how the sum is treated. Thus, the negative binomial model cannot be rejected in a Popperian confrontation with the data.

## A MONTE CARLO APPROACH FOR ESTIMATING THE CHANCE OF SUCCESS IN AN OBSERVER PROGRAM

We can understand the potential for failure caused by ignoring aggregation by asking how likely one is to obtain statistically meaningful data for a given level of observer coverage. This question can be answered using a Monte Carlo method, which proceeds according to the following pseudocode:

---

Pseudocode 4.1

1. Specify the level of observer coverage, $N_{tow}$ per simulation, and the total number of simulations $N_{sim}$, and the negative binomial parameters $m$ and $k$. These are estimated from last year's data. Also specify the criterion "success," $d$, and the value of $t_q$.

---

FIGURE 4.1. (a) The negative binomial (solid) model for the frequency of incidental take accurately predicts the observations (hatched). For ease of presentation, we lumped incidental catches of eight or more together. (b) The residuals (differences between the predicted and observed values) show no pattern.

103

2. On the $j^{th}$ iteration of the simulation, for the $i^{th}$ simulated tow, generate a level of incidental take $C_{ij}$ using Equation 4.7. To do this, first generate the probability of $n$ birds in the by-catch for an individual tow, then calculate the cumulative probability of $n$ or fewer birds being obtained in the by-catch. Next draw a uniform random number between zero and 1, and then see where this random number falls in the cumulative distribution. Repeat this for all $N_{tow}$ tows.

3. Compute the mean

$$M_j = \frac{1}{N_{tow}} \sum_{i=1}^{N_{tow}} C_{ij}$$

and the variance

$$S_j^2 = \frac{1}{N_{tow} - 1} \sum_{i=1}^{N_{tow}} (C_{ij} - M_j)^2$$

on the $j^{th}$ iteration of the simulation.

4. Compute the range, in analogy to Equation 4.4:

$$(\text{Range})_j = 2 \frac{S_j}{\sqrt{N_{tow}}} t_q.$$

5. If $(\text{Range})_j$ is less than the specified range criterion for success, increase the number of successes by 1.

6. Repeat steps 2–5 for $j = 1$ to $N_{sim}$. Estimate the probability of success when $N_{tow}$ tows are observed by dividing the total number of successes by $N_{sim}$.

In the calculations reported below, we used $N_{sim} = 150$, $t_q = 1.645$, and the criterion for success that the range was less than $0.25m$. We can then plot the chance that the observer program will meet its target for reliability as a function of the number of tows observed (Figure 4.2). At sample sizes for which the summarized data predict certain success (i.e., that statistically meaningful data will be obtained with probability equal to 1), the unsummarized data predict disaster! Much higher observer levels are needed because of the aggregated nature of the by-catch data.

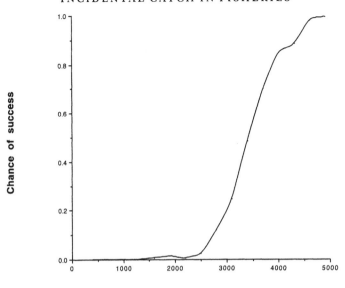

**Number of tows observed**

FIGURE 4.2. Probability of success (defined as the number of simulations in which the reliability target is met) in the observer program as a function of the number of tows observed.

## IMPLICATIONS

Are observer programs doomed to failure? Certainly not—but they should be planned by people who know the data. It was only by studying the unsummarized data that we were able to conclude that the incidental catch was highly aggregated. Bartle (1991) should be commended for inclusion of the raw frequencies, because this allowed us to consider the negative binomial model. Operational and policy recommendations require the best statistical tools, and we must continually confront our models with the actual data (in this case, models of how one presents the data). Only in this manner will the credibility of ecological detectives continue to be enhanced.

# The Confrontation: Sum of Squares

The simplest technique for the confrontation between models and data is the method of the sum of squared deviations, usually called the sum of squares. It has three selling points. First, it is simple; in particular, one need not make any assumptions about the way the uncertainty enters into the process or observation systems. Second, it has a long and successful history in science. It is a proven winner. Third, modern computational methods (Efron and Tibshirani 1991, 1993) allow us to do remarkable calculations associated with the sum of squares. We illustrate the last point in the next chapter while the first two are discussed here.

When starting a project that involves estimation of parameters in a model, we recommend that before you spend lots of time taking measurements and trying to estimate parameters from the data, you simulate data using a Monte Carlo procedure and test your ideas on the simulated data. The advantage is clear. With simulated data, we know the true values of parameters and can thus evaluate how well the procedure works. Often, it turns out that it is difficult to estimate the desired parameters from the kind of data that are being collected—and it is good to know that before you start the empirical work.

## THE BASIC METHOD

To illustrate the method of the sum of squares, we consider a simple model. The observed data consist of independent variables $X_1, \ldots, X_n$ and the dependent variables $Y_1$,

. . . $Y_n$, without observation error. We assume the process model

$$Y_i = A + BX_i + CX_i^2 + W_i, \qquad (5.1)$$

where $W_i$ is the process uncertainty and $A$, $B$, and $C$ are parameters. In general, we might call the parameters $p_0$, $p_1$, and $p_2$; we adopt this notation shortly.

The method is based on a simple idea: for fixed values of the parameters and for each value of the independent variable, we use the process model Equation 5.1 to construct a predicted value of the dependent variable by ignoring the process uncertainty. That is, for a given value of $X$ and estimated $A_{est}$, $B_{est}$, and $C_{est}$ values of the parameters, we predict the value of $Y$ to be

$$Y_{pre,i} = A_{est} + B_{est}X_i + C_{est}X_i^2. \qquad (5.2)$$

Next, we measure the deviation between the $i^{th}$ predicted and observed values by the square $(Y_{pre,i} - Y_{obs,i})^2$. We sum the deviations $(Y_{pre,i} - Y_{obs,i})^2$ over all the data points,

$$\mathcal{S}(A_{est}, B_{est}, C_{est})$$

$$= \sum_{i=1}^{n} (Y_{pre,i} - Y_{obs,i})^2$$

$$= \sum_{i=1}^{n} (A_{est} + B_{est}X_i + C_{est}X_i^2 - Y_{obs,i})^2, \qquad (5.3)$$

to obtain a measure of fit between the model and the data when the parameters are $A_{est}$, $B_{est}$, and $C_{est}$. The sum of squares is a function of the parameters, and the notation $\mathcal{S}(A_{est}, B_{est}, C_{est})$ reminds us that the sum of squares depends on the estimated values of the parameters $A_{est}$, $B_{est}$, and $C_{est}$. The best model is the one with parameters that minimize the sum of squares.

The method is easy to use and makes the fewest assumptions about the data. For example, there are no assumptions made about the nature of the stochastic term in Equation 5.1. The measure of deviation, the square of the difference between the observed and predicted values, has two main advantages. First, if one is attempting to find analytical solutions, then squares are good because the derivatives of $\mathcal{S}(A_{est}, B_{est}, C_{est})$ are easily found. Second, we shall show later (Chapter 7) that if the stochastic term in Equation 5.1 is normally distributed, then the sum of squares is identical to other methods of confrontation. The disadvantage of the squared measure of deviation is that it has an accelerating penalty: a deviation that is twice as large contributes four times as much to the sum of squares. There is no a priori reason to choose such a measure of deviation. However, all the numerical procedures we describe below work as effectively if we replace the measure of deviation $(Y_{pre,i} - Y_{obs,i})^2$ by the absolute deviation $|Y_{pre,i} - Y_{obs,i}|$. Alternatives to the sum of squares are discussed by Rousseeuw (1984). The diligent ecological detective should think carefully about which goodness-of-fit criterion is most appropriate for the problem.

In a case such as Equation 5.3, we can use elementary calculus and remember that a necessary condition for a minimum is that the first derivatives of $\mathcal{S}(A_{est}, B_{est}, C_{est})$ with respect to each of these parameters must vanish at the minimum. Taking the derivatives and setting them equal to zero gives

$$\frac{\partial \mathcal{S}\ (A_{est}, B_{est}, C_{est})}{\partial A_{est}}$$

$$= \sum_{i=1}^{n} 2(A_{est} + B_{est}X_i + C_{est}X_i^2 - Y_{obs,i}) = 0,$$

$$\frac{\partial \mathcal{S}\ (A_{est}, B_{est}, C_{est})}{\partial B_{est}}$$

$$= \sum_{i=1}^{n} 2X_i(A_{est} + B_{est}X_i + C_{est}X_i^2 - Y_{obs,i}) = 0,$$

$$\frac{\partial \mathscr{S} \ (A_{est}, B_{est}, C_{est})}{\partial C_{est}}$$

$$= \sum_{i=1}^{n} 2X_i^2(A_{est} + B_{est}X_i + C_{est}X_i^2 - Y_{obs,i}) = 0. \quad (5.4)$$

These are three linear equations for the three unknowns $A_{est}$, $B_{est}$, and $C_{est}$ that can be solved (even by hand!) with relative ease to obtain values for the parameters.

An alternative to using calculus is to conduct a numerical search over a reasonable range of parameter values [$A_{min}$, $A_{max}$], [$B_{min}$, $B_{max}$], and [$C_{min}$, $C_{max}$] and determine those values that minimize the sum of squares. A pseudocode for conducting a systematic numerical search over the parameter space is:

---

Pseudocode 5.1

1. Input the data $\{X_i, Y_{obs,i}, i = 1, \ldots, n\}$, the range of parameter values ($A_{min}$, $A_{max}$, $B_{min}$, $B_{max}$, $C_{min}$, and $C_{max}$), and the size of the increment used for cycling over the parameters ($Step_A$, $Step_B$, $Step_C$).

2. Systematically search over parameter space from $A_{est} = A_{min}$, $B_{est} = B_{min}$, $C_{est} = C_{min}$ to the maximum values in increments of $Step_A$, $Step_B$, $Step_C$, respectively. For each set of parameter values, initialize $\mathscr{S} = 0$.

3. For each value of the parameters $\{A_{est}, B_{est}, C_{est}\}$, cycle over $i = 1$ to $n$ and increment $\mathscr{S}$ by adding $(Y_{pre,i} - Y_{obs,i})^2$, where $Y_{pre,i}$ is the predicted value of $Y_i$, based on the process model and the value of $X_i$.

4. After cycling over the data, compare the sum of squares $\mathscr{S}$ with the current best value $\mathscr{S}_{min}$. If $\mathscr{S} < \mathscr{S}_{min}$, then set $\mathscr{S}_{min} = \mathscr{S}$ and set $A^* = A_{est}$, $B^* = B_{est}$, and $C^* = C_{est}$.

5. If the maximum values of the parameters have been reached, then stop. Otherwise, return to Step 2 and increase the parameter values.

---

109

The output of either the system of linear equations Equation 5.4 or the numerical search is a set of "best-fit" parameters $A^*$, $B^*$, and $C^*$ (which are the values of the parameters that make the sum of squares the smallest), the predicted values of the dependent variable, and the minimum value $\mathcal{S}_{min} = \mathcal{S}(A^*, B^*, C^*)$.

Recall our starting recommendation to test estimation methods on data that are generated by Monte Carlo methods, so that you know the process that generated the data. A pseudocode to generate data for Equation 5.1 is the following:

---

Pseudocode 5.2

1. Specify values of the parameters $A$, $B$, and $C$, the number of data points to be generated, and the distribution of the process uncertainty. Set $i = 1$.
2. Choose $X_i$ (e.g., by systematic choice of the independent variable $X$).
3. Choose a particular value $w_i$ of the process uncertainty $W_i$.
4. Determine $Y_i$ according to $Y_i = A + BX_i + CX_i^2 + w_i$.
5. Increase $i$ by 1. If this is less than the number of data points to be generated, return to Step 2. Otherwise, stop.

---

To employ this pseudocode, we first specify values for $A$, $B$, and $C$ and a distribution for $W_i$. For example, if the true values of the parameters are $A = 1$, $B = 0.5$, and $C = 0.25$, and $W_i$ is uniformly distributed between $-3$ and $3$, one iteration of a Monte Carlo program based on Pseudocode 5.2 gave:

| $X$ | Deterministic $Y$ | Resulting $Y$ |
|---|---|---|
| 1 | 1.75 | 2.0411 |
| 2 | 3 | 0.4696 |
| 3 | 4.75 | 5.8773 |
| 4 | 7 | 6.0116 |
| 5 | 9.75 | 12.462 |

| X | Deterministic Y | Resulting Y |
|---|---|---|
| 6 | 13 | 14.942 |
| 7 | 16.75 | · 16.994 |
| 8 | 21 | 18.508 |
| 9 | 25.75 | 25.098 |
| 10 | 31 | 31.495 |

In general, we only know the left-hand ($X$) and right-hand (resulting $Y$) columns, and from this information want to estimate parameters and then determine how well the chosen model fits the data. We do this using Pseudocode 5.1. In using the associated program, we let $A$ range from 0 to 3 in steps of 0.1, $B$ from 0 to 2 in steps of 0.05, and $C$ from 0 to 1 in steps of 0.025, and determined $A^* = -0.1$, $B^* = 1.05$, and $C^* = 0.2$. Note that these are not the true parameters. However, we have only ten data points and are trying to determine three parameters, so it would be overly optimistic to expect the method to select the true parameter values. Furthermore, we can now add a "Predicted" column to the table relating $X$ and $Y$ and see that the predictions are actually quite accurate:

| X | Deterministic Y | Resulting Y | Predicted Y |
|---|---|---|---|
| 1 | 1.75 | 2.0411 | 1.15 |
| 2 | 3.00 | 0.4696 | 2.8 |
| 3 | 4.75 | 5.8773 | 4.85 |
| 4 | 7.00 | 6.0116 | 7.30 |
| 5 | 9.75 | 12.462 | 10.15 |
| 6 | 13 | 14.942 | 13.4 |
| 7 | 16.75 | 16.994 | 17.05 |
| 8 | 21 | 18.508 | 21.1 |
| 9 | 25.75 | 25.098 | 25.55 |
| 10 | 31 | 31.495 | 30.4 |

Here the "Predicted $Y$" is determined using the best parameter values.

## GOODNESS-OF-FIT PROFILES

It is often very helpful to consider how sensitive the fit of the model and the data, measured by the sum of squares, is to variation in the parameters. This can be done with the "goodness-of-fit" profile, by systematically varying one parameter and then searching over the others to find the values that minimize the sum of squares. Doing this provides three kinds of information. First, it tells us how the sum of squares behaves if one of the parameters (the one that we systematically vary) is known. Second, it tells us how the values of the best choices of the other parameters (the ones we minimize over) depend on the one that is systematically varied; it gives information concerning how sensitive the parameters are to one another. Third, it gives us some notion of confidence in our estimate of the parameter. If the sum of squares is very flat as we vary a parameter, then we should have little confidence in the "best" estimate. This notion is made much more precise in Chapter 7, when we discuss the likelihood profile.

Denoting the goodness-of-fit profile by $\mathcal{G}$, the model described by Equation 5.1 has three goodness-of-fit profiles:

$$\mathcal{G}(A) = \min_{B_{\text{est}}, C_{\text{est}}}$$

$$\sum_{i=1}^{n} (A_{\text{est}} + B_{\text{est}} X_i + C_{\text{est}} X_i^2 - Y_{\text{obs},i})^2,$$

$$\mathcal{G}(B) = \min_{A_{\text{est}}, C_{\text{est}}}$$

$$\sum_{i=1}^{n} (A_{\text{est}} + B_{\text{est}} X_i + C_{\text{est}} X_i^2 - Y_{\text{obs},i})^2,$$

$$\mathcal{G}(C) = \min_{A_{\text{est}}, B_{\text{est}}}$$

$$\sum_{i=1}^{n} (A_{\text{est}} + B_{\text{est}} X_i + C_{\text{est}} X_i^2 - Y_{\text{obs},i})^2, \quad (5.5)$$

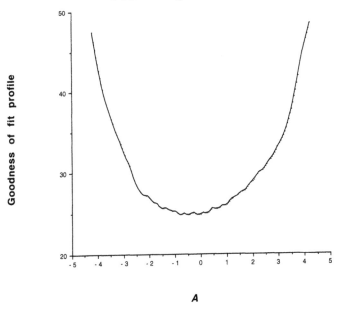

**A**

FIGURE 5.1 The goodness-of-fit profile for the parameter $A$, based on the first equation in Equation 5.5. Values of $A$ in the range from about $-1.0$ to $0.5$ give roughly the same minimum value of the sum of squares. The broad "bowl" in the profile suggests that other values of $A$ are consistent with the data.

where, for example, $\min_{B_{est}, C_{est}}$ means that we find the minimum, as above, over the choices of $B_{est}$ and $C_{est}$. The first goodness-of-fit profile in Equation 5.5 leads to the function $\mathcal{G}(A)$ for optimal values of $B^*(A)$ and $C^*(A)$ as $A$ varies (Figure 5.1). We encourage you to write a pseudocode or program to find it. This is most easily done by thinking about how Pseudocode 5.2 needs to be modified.

In the more general setting, we relate the dependent and independent variables by the process model

$$Y_i = f(X_i, W_i | p_1, p_2, \ldots, p_m), \tag{5.6}$$

where $X$ is the independent variable, $W$ is the process uncertainty (with $X_i$ and $W_i$ indicating the $i^{th}$ values), $\{p_1, p_2, \ldots, p_m\}$ are the parameters, and $f(X_i, W_i | p_1, p_2, \ldots, p_m)$ is the presumed functional relationship between the independent variable, process noise, parameters, and dependent variable. The generalization of Equation 5.3 is

$$\mathcal{S}(p_{1_{est}}, p_{2_{est}}, \ldots, p_{m_{est}})$$

$$= \sum_{i=1}^{n} (Y_{pre,i} - Y_{obs,i})^2$$

$$= \sum_{i=1}^{n} [f(X_i, 0 | p_{1_{est}}, p_{2_{est}}, \ldots, p_{m_{est}}) - Y_{obs,i}]^2. \tag{5.7}$$

Note that we have set $W_i = 0$ in the predicted value of $Y_i$, treating the predicted value as if it were deterministic. Depending on the functional relationship, it may be possible to determine the parameters that make the sum of squares a minimum by taking derivatives, but the numerical method will usually work without any problem. Difficulties arise if there is more than one local minimum (such problems are discussed in Chapter 11).

The generalization of the goodness-of-fit profile for the first parameter is

$$\mathcal{G}(p_{1_{est}}) = \min_{p_{2_{est}}, \ldots, p_{m_{est}}}$$

$$\sum_{i=1}^{n} [f(X_i, 0 | p_{1_{est}}, p_{2_{est}}, \ldots, p_{m_{est}}) - Y_{obs,i}]^2. \tag{5.8}$$

## MODEL SELECTION USING SUM OF SQUARES

Written in the more general framework of Equation 5.6, the model described by Equation 5.1 is

$$Y_i = p_0 + p_1 X_i + p_2 X_i^2 + W_i. \tag{5.9}$$

In the preceding discussion, we generated data according to this particular model and then determined parameters assuming that we knew this model to be correct. Usually, we are not that lucky because we do not know that the model is correct. For example, for the same data generated by the pseudocode, we could envision three models:

Model 1:    $Y_i = p_0 + W_i,$

Model 2:    $Y_i = p_0 + p_1 X_i + W_i,$

Model 3:    $Y_i = p_0 + p_1 X_i + p_2 X_i^2 + W_i.$    (5.10)

Actually, there are even more models, obtained, for example, by including the constant and quadratic terms but not the linear term, or the linear term and the quadratic term but not the constant, but the three in Equation 5.10 are enough.

Suppose that we only knew the data given after Pseudocode 5.2 and confronted those data with model 1, 2, or 3. Using the sum of squares method for model 1, the best-fit parameter is $p_0^* = 13.4$ and the sum of the squared deviations is 913.9. For model 2, the best-fit parameters are $p_0^* = -4.2$ and $p_1^* = 3.2$, and the sum of the squared deviations is 43.33. For model 3, the best-fit parameters were found before and the sum of squared deviations is 24.985. We expect a model with more parameters to fit better in the sense that the sum of squares will be smaller. But we also expect that adding more parameters to a model leads to increasing difficulty of interpretation. Is model 3 actually an improvement over model 2? More generally, how do we compare a model with $m$ parameters to a model with $n$ parameters?

Suppose that a model with $m$ parameters has the sum of squares $SSQ(m)$, which will generally decrease as $m$ increases. However, it makes sense to penalize the introducion of additional parameters. There are a number of ways in which this can be done (Efron and Tibshirani 1993, 242 ff.). The simplest comparison replaces the sum of squares by

$$\frac{SSQ(m)}{n - 2m}.$$    (5.11)

115

Note that we penalize the introduction of each parameter by effectively reducing the number of data points by 2. The criterion in Equation 5.11 is one of a number discussed by Efron and Tibshirani; others include Mallows' $C_p$ (Mallows 1973; Draper and Smith 1981) and the Bayesian information criterion. The criterion in Equation 5.11 and Mallow's $C_p$ are approximately the same (Efron and Tibshirani 1993; also see Nishi 1984; Hongzhi 1989).

For the data generated by the pseudocode, the criterion in Equation 5.11 is 114.2, 7.22, and 6.25 for models 1, 2, and 3, respectively. We thus choose model 3 as the best predictor of the dependent variable. This is somewhat comforting, because model 3 is "correct" in the sense that it was used to generate the data. On the other hand, when we do not know the process model, a simpler model may fit the data better than a more complex model—even if the more complex model is "more realistic" or if the simpler model is biologically incorrect. This is a tough fact of life, but one that must constantly be considered by ecological detectives, who must pay attention to both statistical and biological considerations.

In general, we might ask how the preferred model would act with other data sets. The problem is that often we do not have such other data sets. Here the bootstrap method, described in Chapter 3, is useful. We use the bootstrap method to generate additional data sets and then compare various models, using the criterion in Equation 5.11. This kind of comparison gives us a sense of how confident we should be with the model that wins the competition arbitrated by the actual data set. The comparison also brings us closer to a Bayesian/Lakatosian viewpoint. Almost all ecological models can be built with differing levels of complexity; it is easy to add additional parameters. Since one of the principal tasks of the ecological detective is to consider the support the data provide for alternative models, we need a way of comparing models, and the sum of squares is such a method. However, when we want to chose a "best" model, then Equation 5.11 or other criteria such as Mallows' $C_p$ can be used. The choice of a best model implies that in some way we reject the others and accept the best one. A Bayesian would, instead, want to assign relative degrees of belief to the competing models. The comparison of models with bootstrap data sets lets us mimic the Bayesian approach. It is exactly that kind of competition that we now consider.

# The Evolutionary Ecology of Insect Oviposition Behavior

## MOTIVATION

The study of clutch size was formally initiated by David Lack about fifty years ago (Lack 1946, 1947, 1948), and continues to be a major field of interest, involving both theoretical and empirical aspects (e.g., Godfray et al. 1991; Mangel et al. 1994). Although Lack was interested in birds, his ideas have been applied widely; here we consider the oviposition behavior of insect parasitoids. These insects, usually Hymenoptera (wasps), have a typical life history pattern in which adults are free ranging, often able to fly great distances, and lay their eggs in or on the eggs, larvae, pupae, or adults of other insects. The eggs hatch and the larvae consume the host and pupate. Often more than one egg is laid in a host. The size of the clutch laid in a host can affect both the probability that offspring will emerge and the size of the offspring, which is usually related to parental fecundity.

In this chapter we confront two kinds of models for oviposition behavior with the data. We illustrate how the sum of squares and bootstrap methods can be used to select between models when the process and/or observation uncertainties are not known.

## THE ECOLOGICAL SETTING

Armored scales are pests of fruit trees worldwide. Parasitic wasps in the genus *Aphytis* are used for biological control of armored scales, which are hosts for the parasitoid. When a female encounters a scale insect, she places eggs under the armor of the scale insect, on the soft part of its body. The time needed to get under the armor and place an egg is the handling time. Rosenheim and Rosen (1991) studied how the number of eggs laid by female *Aphytis linganensis* in their first encounter with a host depended on the number of eggs she carried (egg complement). The number of eggs was manipulated by choosing pupae that emerged from host scale insects of varying size, and by raising females under different temperature regimes to vary egg maturation rates. Rosenheim and Rosen minimized the variation in host size, but it could not be eliminated. However, host size varied independently of egg complement (J. Rosenheim, personal communication).

## THE DATA

The main data are the number of clutches of different sizes laid by insects with different egg complements (Table 6.1). There are 102 such combinations of egg complement $E$ and clutch size $C$; we denote the $i^{th}$ data pair by $\{E_i, C_i\}$; here $i = 1$ to $N_c = 102$. In summarized form, we use the number $N(E,C)$ of clutches of size $C$ when the egg complement is $E$; here $E = 1$ to 23 [although some values of $N(E,C)$ will be 0 because there were no observations at some levels of egg complement] and $C = 1$ to 4. We will not allow clutches larger than 4, since none were observed.

Rosenheim and Rosen found that the size of females who emerged from hosts depended upon the number of eggs laid in that host. Since the fecundity of a female depends upon her size, we can compute the potential number of grandchildren from the daughters of a female who lays a clutch of a given size in a host (Figure 6.1).

119

TABLE 6.1. Number of eggs laid in first encounters with host scale insects by *Aphytis* parasitoids with different egg complements. (Nonsummarized data were kindly provided by Jay Rosenheim.)

| Egg complement | Number of observations of clutch size | | | |
|---|---|---|---|---|
| | 1 | 2 | 3 | 4 |
| 4 | 0 | 2 | 1 | 0 |
| 5 | 0 | 5 | 1 | 0 |
| 6 | 1 | 11 | 3 | 0 |
| 7 | 1 | 5 | 1 | 0 |
| 8 | 0 | 2 | 1 | 0 |
| 9 | 0 | 1 | 0 | 1 |
| 10 | 0 | 4 | 3 | 0 |
| 11 | 0 | 3 | 4 | 0 |
| 12 | 0 | 1 | 6 | 0 |
| 13 | 1 | 2 | 4 | 0 |
| 14 | 0 | 0 | 3 | 1 |
| 15 | 0 | 3 | 4 | 0 |
| 16 | 0 | 2 | 6 | 0 |
| 17 | 0 | 2 | 4 | 0 |
| 18 | 0 | 0 | 2 | 0 |
| 19 | 0 | 0 | 6 | 0 |
| 20 | 0 | 0 | 2 | 0 |
| 21 | 0 | 1 | 1 | 0 |
| 22 | 0 | 0 | 0 | 0 |
| 23 | 0 | 0 | 1 | 0 |

## THE MODELS

We discuss three kinds of models that are used to apply Lack's ideas to insect oviposition behavior (also see Rosenheim and Rosen 1991).

*Single-Host Models.* If a reasonable measure of fitness is the number of grandchildren produced by the offspring that emerge from a host (Charnov and Skinner 1984), then one possibility is that clutch size has evolved to maximize this number, which is called the single-host maximum clutch (Mangel 1987) and abbreviated as the SHM clutch.

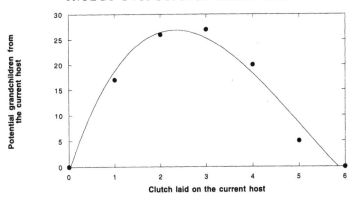

FIGURE 6.1. The data of Rosenheim and Rosen (1991) lead to a domed relationship between the number of eggs laid in a host by a female and the total number of eggs (potential grandchilden) that her daughters will have. From the perspective of a single host, the optimum clutch is 3, which produces 27 potential grandchildren, although a clutch of 2 produces nearly as many. The regression equation $y = -0.857 + 26.5x - 7.43x^2 + 0.5x^3$ provides a very good fit to the data.

In such a case, if host size is constant, clutch size will be constant across different egg complements. If host size varies, then clutch may depend upon the size (or another measure of host quality) but not on egg complement.

*Rate-Maximizing Models.* On the other hand, laying additional eggs in hosts prevents the parasitoid from searching for new hosts. Thus, the appropriate measure of fitness might be the rate at which grandchildren are accrued from oviposition (Charnov and Skinner 1984). This is an application of optimal foraging theory (Stephens and Krebs 1986) to oviposition behavior. The rate of accumulation of reproductive success is determined by (i) reproductive success from the host, (ii) the handling time associated with the clutch, and (iii) the travel time between hosts. The prediction is that on the first encounter, clutch size will be the same across different egg complements (again assuming that host quality is constant).

The difference between these two models can be summarized as follows. The SHM clutch is the one that produces the largest number of potential grandchildren from a single host. However, note from Figure 6.1 that the first egg in a clutch produces proportionately a greater number of potential grandchildren (17) (at the margin) than the second or third eggs (9 or 1 potential grandchildren). The rate-maximizing clutch balances the decreasing increments in potential grandchildren per egg laid against the time to find the next host (Charnov and Skinner 1984). As search time decreases, the rate-maximizing clutch also decreases. However, both the SHM and rate-maximizing clutches are independent of egg complement.

*State-Variable Models.* When a parasitoid begins life with a limited number of eggs and matures only a few eggs during the course of her life, the problem becomes more complicated, because an egg used now cannot be used later—there is a tradeoff between current and future reproduction. In such a case, one must compute the lifetime reproductive success of the parasitoid, taking changes of egg complement into account.

To deal with a physiological variable such as egg complement, state-variable models (Mangel 1987; Mangel and Clark 1988) are required. The details of such models are beyond presentation here but we encourage you to consult some of the primary literature (see Mangel 1987; Mangel and Clark 1988; Mangel and Ludwig 1992). However, the important general prediction from such models is that clutch size should increase as egg complement increases. The general reason for this prediction is the tradeoff between current and future reproduction. An egg used now is clearly not available later, and there is a decreasing payoff from additional eggs in hosts (Figure 6.1). Thus, when egg complement is high, we predict that the female may lay close to a SHM clutch, but when egg complement is smaller the clutch may be considerably smaller than the SHM clutch.

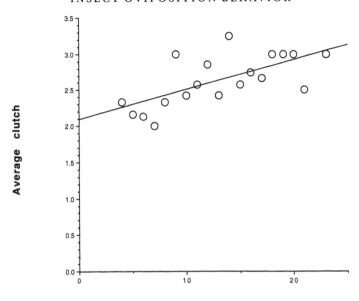

Egg complement

FIGURE 6.2. The average clutch increases with egg complement. The regression equation $Y = 2.089 + 0.0415X$ explains only about 45% of the variation.

In summary, we have two kinds of models. Single-host maximum or rate-maximizing models predict a clutch that is independent of egg complement. State-variable models predict that clutch size will increase with egg complement.

## THE CONFRONTATION

The simplest confrontation is to average the data in Table 6.1 and ask if the trend is for clutch size to increase with egg complement. The answer is yes (Figure 6.2), but there is considerable unexplained variation. Our goal is to develop an understanding beyond Figure 6.2. Because many of the sources of uncertainty in this experiment are unknown (they include genotypic effects, maternal effects, individual experience, etc.), and the probability distributions for these uncertainties are unknown, we use the sum of squares to compare the different models.

### The Data Themselves

We begin with the egg-independent clutches. Suppose that the fixed clutch is $c_f$. The sum of squares for this fixed clutch is

$$SSQ(c_f) = \sum_{E=1}^{23} \sum_{C=1}^{4} (C - c_f)^2 \, N(E,C).$$
(6.1)

That is, we weight the sum of squared deviations by the number of observations at that egg complement and clutch. There are four values of the fixed clutch ($c_f = 1, 2, 3,$ or $4$), each of which produces a different value of $SSQ(c_f)$. A pseudocode to do this is:

---

Pseudocode 6.1
1. Read the data from Table 6.1 into a table called $N(E,C)$.
2. Cycle over $c_f$.    Set $SSQ(c_f) = 0$.
3. Cycle over $E = 4$ to $23$.
4. Cycle over $C = 1$ to $4$.
5. Replace $SSQ(c_f)$ by $SSQ(c_f) + (C - c_f)^2 \, N(E,C)$. Return to step 3.

---

Because there are no estimated parameters in Equation 6.1, the appropriate measure for comparing the sum of squares is $SSQ(c_f)/N_c$. Using this pseudocode leads to

| Fixed clutch, $c_f$ | $SSQ(c_f)/N_c$ |
|:---:|:---:|
| 1 | 2.686 |
| 2 | 0.627 |
| 3 | 0.569 |
| 4 | 2.510 |

From these results, we conclude that models with fixed clutches of 2 or 3 are approximately equally good predictors (we shall be more specific about this later), and that models with fixed clutches of 1 or 4 are removed from the competition between the models. These results are consonant with what we might conclude from Figure 6.2.

The output of a dynamic state-variable model is typically one where the clutch the parasitoid lays increases with egg complement $e$; hence we denote clutch size when egg complement is $e$ by $c(e)$. Adopting this viewpoint, the simplest variable-clutch model is one in which the parasitoid switches from clutch size $c_1$ to clutch size $c_2 > c_1$ at egg complement $e_1$. Then

$$c(e) = \begin{cases} c_1 & \text{if } e \leq e_1, \\ c_2 & \text{if } e > e_1. \end{cases} \tag{6.2}$$

The sum of squares for this model is

SSQ(single switch)

$$= \sum_{E=1}^{23} \sum_{C=1}^{4} [C - c(E)]^2 \, N(E,C), \tag{6.3}$$

and there are three parameters that are estimated from the data ($c_1$, $c_2$, and $e_1$). A pseudocode to compute the sum of squares in Equation 6.3 is:

---

Pseudocode 6.2

1. Read the data from Table 6.1 into the table $N(E,C)$. Set the minimum sum of squares SSQ* $= 10^6$ (or any other large value).

2. Cycle over values of $c_1$, $c_2$, and $e_1$. For each combination set SSQ(single switch) $= 0$.

3. For each combination of $c_1$, $c_2$, and $e_1$, cycle over $E = 1$ to 23 and $C = 1$ to 4.

4. If $E \leq e_1$ then replace SSQ(single switch) by SSQ(single

125

switch) $+ (C - c_1)^2 N(E,C)$. Otherwise, replace it by
SSQ(single switch) $+ (C - c_2)^2 N(E,C)$.

5. After cycling over all values of $E$ and $C$, compare SSQ*
   with SSQ(single switch). If the newly computed
   SSQ(single switch) $<$ SSQ*, then replace SSQ* with
   it, and set $c_1^* = c_1$, $c_2^* = c_2$, and $e_1^* = e_1$. Return to
   step 3.

---

The output of this pseudocode gives the minimum sum of
squared deviations, and the best values of the parameters
(i.e., those values that make the sum of squared deviations
the smallest). Since there are three parameters, the appropriate measure for model selection is SSQ*$/N_c - 6$. A program using this pseudocode gives $c_1^* = 2$, $c_2^* = 3$, and $e_1^*$
$= 8$. That is, we predict the parasitoids will lay clutches of
size 2 if they have 8 or fewer eggs, and clutches of size 3
otherwise. The measure for model comparison is SSQ*$/N_c$
$- 6 = 0.354$. We thus conclude that the variable-clutch
model with a single switch point outcompetes any of the
fixed-clutch models.

A variable-clutch model with two switch points is

$$c(e) = \begin{cases} c_1 & \text{if } e \le e_1, \\ c_2 & \text{if } e_1 < e \le e_2, \\ c_3 & \text{if } e > e_2, \end{cases} \tag{6.4}$$

and has five parameters ($c_1$, $c_2$, $c_3$, $e_1$ and $e_2$). The sum of
squared deviations is computed in analogy to Equation 6.3
and by a similar pseudocode. Because there are five parameters, the measure for model comparison is SSQ*$/N_c - 10$,
which has the value 0.369. Note that the two-switch-point
model with more parameters actually has a poorer performance measure than the single-switch-point model with
fewer parameters. The reason is the penalty associated with
the number of parameters (see Equation 5.11 and
following).

Tentative conclusions are that among fixed-clutch

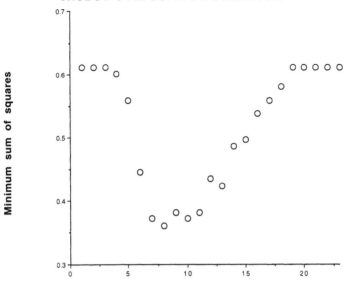

**Switching value of egg complement**

FIGURE 6.3. The goodness-of-fit profile for the switching value $e_1$ of egg load in the single-switch state-variable model. The sum of squares is minimized over the choices of $c_1$ and $c_2$. Note that values of $e_1$ in the range 7–11 give approximately the same minimum value.

models, those with fixed clutches of 2 or 3 are reasonable competitors, and that a variable-clutch model with a single switch point outcompetes either of the fixed-clutch models.

Since the variable-clutch model involves three parameters, we can compute goodness-of-fit profiles. We will focus on the switch value $e_1$. The goodness-of-fit profile is

$$\mathscr{G}(e_1) \;=\; \min_{c_1, c_2} \sum_{E=1}^{23} \sum_{C=1}^{4} \left[ C - c(E) \right]^2 N(E, C),$$

$$(6.5)$$

where $c(E)$ is given by Equation 6.3. Each fixed choice of $e_1$ generates optimal values $c_1^*(e_1)$ and $c_2^*(e_1)$. The goodness-of-fit profile (Figure 6.3) shows that values of $e_1$ in the range of 7 to 11 all give approximately the same value for the min-

127

imum sum of squares. One of the weaknesses of the sum of squares method is that it is difficult to attribute confidence, in the form of probability statements, to the value of the sum of squares (in large part because we make no assumptions about the form of the uncertainty). Likelihood and Bayesian methods, discussed in later chapters, allow us to do this.

### A Bootstrap Competition

How much better is the variable-clutch model? For example, would it win competitions arbitrated by other data sets? To answer this question, we turn to a bootstrap competition between the models. The notion is to resample the data a large number of times (we used 10 000) and compare the models using the sum of squares criterion for each resampled data set.

The first step involves generating bootstrap samples. We use the $N_c = 102$ original data points $\{E_i, C_i\}$. A bootstrap data set will have 102 different points, which we denote by $\{E_{bs_k}, C_{bs_k}\}$.

Similarly, we will need to track the total number $N_{bs}(E_{bs_k}, C_{bs_k})$ of clutches of size $C_{bs_k}$ when the egg complement is $E_{bs_k}$. A pseudocode for generating a bootstrap data set is:

---

Pseudocode 6.3
1. Read in the original data $\{E_i, C_i\}$ for $i = 1$ to 102.
2. Cycle over $k$ from 1 to 102. For each value of $k$, randomly choose $j$ between 1 and 102. Set $E_{bs_k} = E_j$ and $C_{bs_k} = C_j$. Cycle over $E = 1$ to 23 and $C = 1$ to 4. If $E = E_{bs_k}$ and $C = C_{bs_k}$, then increase $N_{bs}(E,C)$ by 1.

---

Given a bootstrap data set, we compute the sum of squares for the two fixed-clutch models ($c_f = 2$ or 3) using Equation 6.1 and the sum of squares for the variable-clutch model using Equation 6.3. To allow a given benefit to the egg-independent models, when comparing the models, we used $SSQ(c_f)/N_c$ and $SSQ(\text{single switch})/N_c - 6$ in model comparisons. For a given bootstrap data set, there are two relevant questions. First, in a competition between the fixed-clutch models, which model has the smaller $SSQ(c_f)/N_c$? Second, in a comparison between all three models, which has the smallest sum of squares divided by the adjusted number of parameters?

For 10 000 bootstrap competitions, when the two fixed-clutch models were compared, the model with $c_f = 2$ won 3349 of the competitions and the model with $c_f = 3$ won 6651 of the competitions. Clearly, these results do not provide any firm criterion to confidently select one over the other. However, if you had to predict what the next wasp would do, picking a clutch of size 3 makes more sense.

When the three models were compared, the model with $c_f = 2$ won 19 competitions, the model with $c_f = 3$ won 2 competitions, and the variable-clutch model won 9971 of the competitions. It is tempting to associate probabilities of 19/10 000 or 2/10 000 with the fixed-clutch models, which is approximately right (Efron and Tibshirani 1991, 1993). On the other hand, we conclude that the evidence overwhelmingly supports a variable-clutch model.

## IMPLICATIONS

Given the data (Table 6.1, Figure 6.2), one might argue a priori that any fixed-clutch model must be less likely than the variable-clutch models, so that it is optimistic to use a model that is independent of egg complement. In fact, a reviewer of this book remarked that "no intelligent person would use a fixed clutch model." However, such models— especially rate-maximizing models—have been used so frequently in the analysis of problems in behavioral ecology (e.g., Stephens and Krebs 1986) that many people now expect organisms to maximize rate and are surprised when it does not happen (e.g., Cronin and Strong 1993; Rosenheim and Mangel 1994). This tradition has evolved in part because rate-maximizing models are easy to use and in part because for so many years there were no feasible alternatives. Our analysis suggests that dynamic state-variable models are exceedingly more likely than egg-independent models.

Our conclusions are similar to those reached by Rosenheim and Rosen using logistic regression (Hosmer and Lemeshow 1989; Collett 1991). Both logistic regression and the bootstrap methods that we used assume that the data are "representative" of the natural world. This concern usually arises when one creates bootstrap samples, but the same is true for logistic regression. The tradeoff is this. When using the bootstrap methods for model comparison, we make no assumptions about how uncertainty enters the system. The disadvantage is that we are unable to make accurate probability statements. When using logistic regression, we make distributional assumptions about the uncertainty and from these we are able to make probability statements. The disadvantage is that the assumption is made and control of the analysis is turned over to the computer, rather than having the ecological detective at the helm.

In this chapter, because we ignored the details of how uncertainty enters into the behavioral processes, we used the sum of squares to compare essentially qualitative predictions of models. In subsequent chapters, we make more assumptions about the nature of the uncertainty and are able to make more precise quantitative comparisons.

# The Confrontation: Likelihood and Maximum Likelihood

## OVERVIEW

The method of sum of squares can be used to find the best fit of a model to the data under minimal assumptions about the sources of uncertainty. Furthermore, goodness-of-fit profiles and bootstrap resampling of the data sets allow us to make additional inferences about the competition between different models. All of this can be done without assumptions about how uncertainty enters into the system. However, there are many cases in which the form of the probability distributions of the uncertain terms can be justified. For example, if the deviations of the data from the average very closely follow a normal distribution, then it makes sense to assume that the sources of uncertainty are normally distributed.

In such cases, we can go beyond the sum of squares and use the methods of maximum likelihood, which are discussed in this chapter. The likelihood methods discussed here allow us to calculate confidence bounds on parameters (something we could not do with the sum of squares), and to test hypotheses in the traditional manner. In addition, likelihood forms the foundation for Bayesian analysis, which is discussed in Chapter 9.

In this chapter, we use the probability distributions discussed in Chapter 3 to (i) find parameters of a given model that provide the best fit to the data (called maximum likelihood estimation), (ii) compare alternative hypotheses (by using the likelihood ratio test or its generalization to non-nested models), and (iii) calculate confidence bounds (using the method of the likelihood profile). We now introduce these methods.

## LIKELIHOOD AND MAXIMUM LIKELIHOOD

For any of the probability distributions considered in Chapter 3, the probability of observing data $Y_i$, given a particular parameter value $p$, is

$$\Pr\{Y_i | p\}. \tag{7.1}$$

The subscript on $Y_i$ indicates that there are many possible outcomes (for example, $i = 1, 2, \ldots I$), but only one parameter $p$. For example, suppose that $Y_i$ follows a Poisson distribution with rate parameter $r$. Then in one unit of time we predict that $Y_i = k$ with probability

$$\Pr\{Y_i = k \mid \text{rate parameter} = r\} = \frac{e^{-r}r^k}{k!}. \tag{7.2}$$

This expression is also the probability of the "data" given the "hypothesis," where the "data" are $k$ events in one unit of time and the "hypothesis" is that the rate parameter is $r$. When confronting models with data, we usually want to know how well the data support the alternative hypotheses. That is, after collection, the data are known but the hypotheses are still unknown. We ask, "Given these data, how likely are the possible hypotheses?"

To do this, we introduce a new symbol to denote the "likelihood" of the data given the hypothesis:

$$\mathscr{L}\{\text{data} \mid \text{hypothesis}\} \quad \text{or} \quad \mathscr{L}\{Y | p_m\}. \tag{7.3}$$

Note the subtle shift in going from Equation 7.1 to Equation 7.3: $Y$ has no subscript because there is only one observation, but now the parameter is subscripted because there are alternative parameters (hypotheses); for example, we might have $m = 1, 2, \ldots, M$.

The key to the distinction between likelihood and probability is that with probability the hypothesis is known and the data are unknown, whereas with likelihood the data are known and the hypotheses unknown. In general, we assume that the likelihood of the data, given the hypothesis, is proportional to the probability Equation 7.1 (Edwards 1992), so the likelihood of parameter $p_m$, given the data $Y$, is

$$\mathscr{L}\{Y|p_m\} = c \ \Pr\{Y|p_m\}. \tag{7.4}$$

Also, in general, we are concerned with relative likelihoods because we mostly want to know how much more likely one set of hypotheses is relative to another set of hypotheses. In such a case, the value of the constant $c$ is irrelevant and we set $c = 1$. Then the likelihood of the data, given the hypothesis, is equal to the probability of the data, given the hypothesis. Note that although it must be true that if the parameter $p$ is fixed $\Sigma_{i=1}^{I} \Pr\{Y_i|p\} = 1$, when the data $Y$ are fixed, the sum over the possible parameters $\Sigma_{m=1}^{M} \mathscr{L}\{Y|p_m\}$ need not even be finite, let alone equal to 1. It may be helpful to think of likelihood as a kind of unnormalized probability.

For example, suppose that the data were $k = 4$ events in one unit of time. For the Poisson model, Equation 7.2, the likelihood is

$$\mathscr{L}\{4|r\} = \frac{e^{-r} r^4}{4!}. \tag{7.5}$$

If the data were six events in one unit of time, then

$$\mathscr{L}\{6|r\} = \frac{e^{-r} r^6}{6!}. \tag{7.6}$$

By plotting the likelihood as function of $r$ (Figure 7.1a), we get a sense of the range of parameters for which the observations are probable. When looking at this figure, remember that the comparisons are within a particular value of the data and not between different values of the data. For

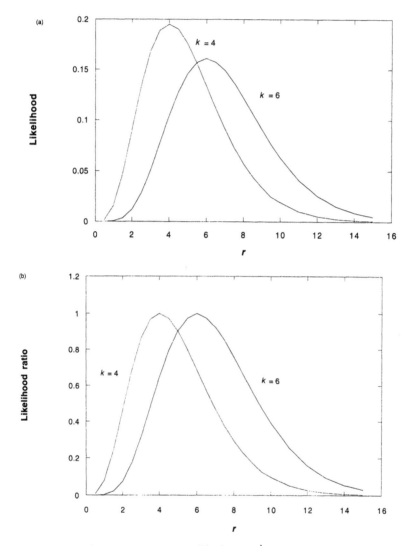

FIGURE 7.1. (a) The likelihood $\mathscr{L}\{k|r\} = e^{-r}r^k/k!$ for $k = 4$ and 6. (b) The likelihood ratio $\mathscr{L}\{k|r\}/\mathscr{L}\{k|r^*\}$, where $r^*$ is the value of the parameter that maximizes the likelihood, for $k = 4$ and 6. (c) The negative log-likelihoods.

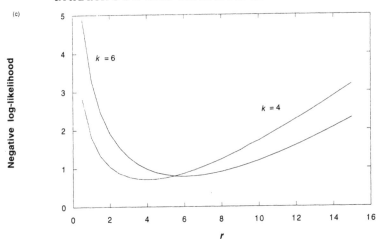

this reason, it is often helpful to scale the likelihoods relative to the parameter value that makes the likelihood as large as possible (Figure 7.1b). For example, when $k = 4$, we see that the most likely value of the parameter is $r = 4$, and that values of $r$ in the range [2,7] are at least half as likely as the most likely parameter. Similarly, when $k = 6$, the most likely value of the parameter is 6 and values of $r$ in the range [4,10] are at least half as likely as the most likely parameter. The parameter that makes the likelihood as large as possible is called the *maximum likelihood estimate (MLE)*.

Because likelihoods may be very small numbers, the tradition is to use the logarithm of the likelihood, called the log-likelihood, for comparisons. This is also called the *support function* (Edwards 1992).

In analogy to the sum of squares, we use the negative of the logarithm of the likelihood, so that the most likely value of the parameter is the one that makes the negative log-likelihood as small as possible:

135

**L**{data | hypothesis}

$$= - \log(\mathcal{L}\{\text{data | hypothesis}\}). \quad (7.7)$$

Then the hypothesis with the most "support from the data" has the smallest value of **L**{data | hypothesis}. For the case we are considering (Figure 7.1c), the maximum likelihood values are $r^* = 4$ and $r^* = 6$ for $k = 4$ and 6, respectively, and it can be seen that these make the negative log-likelihood as small as possible. Thus, we can use the likelihood to decide which hypothesis is most consistent with the data. Schnute and Groot (1992) give a nice summary of inference based on the negative log-likelihood function.

### Multiple Observations

We often have multiple observations of different types of data. Since likelihoods are determined from probabilities, the likelihood of a set of independent observations is the product of the likelihoods of the individual observations. Thus,

$$\mathcal{L}\{Y_1, Y_2, Y_3 | p\} = \mathcal{L}\{Y_1 | p\} \, \mathcal{L}\{Y_2 | p\} \, \mathcal{L}\{Y_3 | p\}, \quad (7.8)$$

and since logarithms are additive, the negative log-likelihoods add:

$$\mathbf{L}\{Y_1, Y_2, Y_3 | p\} = \mathbf{L}\{Y_1 | p\} + \mathbf{L}\{Y_2 | p\} + \mathbf{L}\{Y_3 | p\}. \quad (7.9)$$

Thus, likelihood allows the inclusion of different types of information in a single framework. If a model predicts several different types of observations, we can use likelihood to determine the extent to which the model is consistent with all of the observations.

### Maximum Likelihood and Sum of Squares May Be the Same

One interesting feature of the normal distribution is that the negative log-likelihood and the sum of squares will be minimized at the same values of the parameters. To see this, we begin with the likelihood for $n$ observations $\{Y_i\}$ which follow a normal distribution with mean $m$ and variance $\sigma^2$:

$$\mathcal{L}\{Y|m,\sigma\} = \prod_{i=1}^{n} \frac{1}{\sigma\sqrt{2\pi}} \exp\left(-\frac{(Y_i - m)^2}{2\sigma^2}\right).$$
(7.10)

The negative log-likelihood is

$\mathbf{L}\{Y|m,\sigma\}$

$$= n[\log(\sigma) + \frac{1}{2}\log(2\pi)] + \sum_{i=1}^{n} \frac{(Y_i - m)^2}{2\sigma^2}.$$
(7.11)

To find the value of $m$ that minimizes $\mathbf{L}$, notice that $n[\log(\sigma) + (1/2)\log(2\pi)]$ does not depend on $m$. Therefore, the value of $m$ that minimizes the negative log-likelihood will be one that minimizes the sum on the right-hand side, which is the square deviation between the predicted ($m$) and observed ($Y_i$) values. Many of the familiar problems in regression and analysis of variance assume normal distributions, and therefore the estimated parameters will be the same using sum of squares or maximum likelihood.

### Calculating Averages Using Maximum Likelihood

As an easy introduction to how to use maximum likelihood, let us consider the following set of data. Suppose that the heights (in cm) of ten people are 171, 168, 180, 190, 169, 172, 162, 181, 181, and 177. Also assume that we know that height is normally distributed with standard deviation 10 cm. Therefore the likelihood of any individual height $Y$, if the true mean of the population is $m$, is

$$\mathcal{L}\{Y|m\} = \frac{1}{10\sqrt{2\pi}} \exp\left(-\frac{(Y - m)^2}{200}\right),$$
(7.12)

and the negative log-likelihood for 10 of the ten heights is

$\mathbf{L}\{Y|m\}$

$$= n[\log(10) + \frac{1}{2}\log(2\pi)] + \sum_{i=1}^{n} \frac{(Y_i - m)^2}{200}.$$
(7.13)

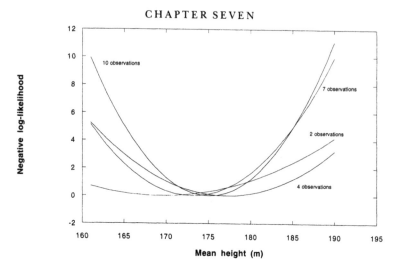

FIGURE 7.2. The negative log-likelihood (scaled so that the minimum is at 0) for the average height of the population when 2, 4, 7, or 10 observations are used.

Figure 7.2 shows the negative log-likelihood for different values of *m* for the data set using the first 2, 4, 7, and finally all 10 observations. In all cases, the minimum **L** has been subtracted from the **L** so that they are all plotted with 0.0 as a minimum. When we use only two data points, the curve is very flat, that is, the alternative hypotheses about *m* have similar likelihoods. As the number of data points used increases, the negative log-likelihood becomes steeper, which indicates that we have more confidence in our knowledge about *m*. Later in this chapter, we show how to find confidence intervals from **L**.

## DETERMINING THE APPROPRIATE LIKELIHOOD

At this point, you may ask, "Given data and hypotheses, what likelihood function should I use?" If you find yourself in this position, then you have not completely specified the model. In particular, you may have a deterministic model but not a stochastic one, because a fully specified stochastic model contains a hypothesis about the way in which randomness enters into the system. If you have not done so, you should return to Chapter 3 and formulate hypotheses about the stochastic components of your model. Ask questions such as: Is there process uncertainty? If so, what kind of distribution is appropriate? Is there observation uncertainty? If so, what kind of distribution is appropriate?

This choice is often made on first principles from the basic distributions described in Chapter 3. For instance, when dealing with simple proportions, the binomial distribution naturally might occur. Data that fall into several possible categories can be described by a multinomial distribution. Counts of rare events could be Poisson or negatively binomially distributed. Quantities that result from the sum of events are often normally distributed, and quantities that result from a series of multiplicative probabilities frequently are log-normal.

You may be able to use the data to distinguish between different probability models for the stochasticity in your system. Different probability models can be thought of as competing hypotheses in exactly the same way that different parameter values are competing hypotheses. Remember that the model consists not only of the deterministic equations, but also of the assumptions about randomness. More simply, examine the residuals, as we did in Chapter 4, to see if there is a systematic pattern to the difference between the model and the data. For example, if the residuals are symmetrically distributed, the normal distribution may be appropriate, but strong skewness in residuals suggests a log-normal distribution.

### Observation and Process Uncertainty

To illustrate the distinction between observation and process uncertainty, imagine a population growth process. If there is only observation uncertainty, then the population dynamics (births and deaths) will be deterministic, but we are unable to accurately estimate population abundance.

Observation uncertainty does not propagate in time. If we underestimate the population in one year, it does not affect the population the next year (the organisms do not know if we overcount or undercount them). As long as our observation uncertainties are independent from year to year, we will be just as likely to overestimate or underestimate the population next year.

If we have process uncertainty but not observation uncertainty, then we estimate population size perfectly (as in many laboratory populations), but the processes of birth and death have random components. If the process uncertainty reduces population size in one year (due to poor births or survival), then the population will be smaller the next year; process uncertainty will propagate over time.

Suppose that we observe a system in which the variable $Y$ depends linearly on the independent variable $X$. We might begin by writing

$$Y = p_0 + p_1 X + W. \tag{7.14}$$

In this equation, $p_0$ and $p_1$ are the parameters to be determined from the data, and $W$ is the process uncertainty (for simplicity, we will not subscript the variables by time or observation number in this section). Now let us explicitly recognize that the observed values of the independent and dependent variables, $X_{obs}$ and $Y_{obs}$, respectively, also involve observation uncertainty by writing

$$Y_{obs} = Y + V_1,$$
$$X_{obs} = X + V_2, \tag{7.15}$$

where $V_1$ and $V_2$ are the observation uncertainties. We combine Equations 7.14 and 7.15 as

$$Y_{obs} = p_0 + p_1 X + W + V_1$$
$$= p_0 + p_1(X_{obs} - V_2) + W + V_1$$
$$= p_0 + p_1 X_{obs} + Z, \tag{7.16}$$

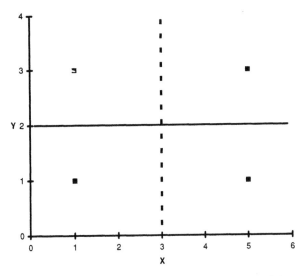

FIGURE 7.3. The four "observations" represent a possible set of data relating $Y$ to $X$. The horizontal line is the interpretation if we believe that $X$ is measured perfectly but that there is process uncertainty. The vertical line is the intepretation if we believe that there is no process uncertainty, but that $X$ is measured imperfectly.

where $Z = W + V_1 - p_1 V_2$ is the "total uncertainty." This is the regression equation usually encountered in statistics books, where it is typically assumed that $X$ is observed perfectly and that $Y$ is subject to process uncertainty.

Why should one think about the sources of uncertainty, particularly to separate process and observation uncertainty, when it is possible to use the last line of Equation 7.16 and ignore the issue entirely? Schnute (1987) illustrates the importance of thinking about the sources of uncertainty. Suppose we have four measurements (Figure 7.3). If we believe that there is no observation uncertainty ($V_1 = V_2 = 0$) but only process uncertainty, then the horizontal line is the appropriate interpretation of the data. In such a case, we assert that $Y$ is independent of $X$, but because of process uncertainty we observe different values for $Y$ at different values

141

of $X$. On the other hand, if we believe that the only uncertainty occurs with the observation of $X$ ($V_1 = W = 0$), then the vertical line is the interpretation. We then assert that $X$ is constant, but measured with uncertainty.

This example, of course, is contrived and most of us would not attempt to draw many conclusions from four data points, especially if they looked like the ones in Figure 7.3. On the other hand, the example does show how our *interpretation* of the data depends on our belief about how randomness is represented in the data. In any comparison of models, the results depend not only on what is actually in the data, but also upon how we believe uncertainty enters into the data. It is always better to recognize such limitations from the outset.

When only observation or process uncertainty is present, we can estimate the amount of variation from the data. For example, in a standard linear regression (Equation 7.16) we assume no observation uncertainty and usually estimate the slope, the intercept, and the variance of the process uncertainty. However, when $X$ is measured imprecisely, it is impossible to estimate the variances of both the observation and process uncertainties. In particular, if both observation and process uncertainty are present, we must either specify the variance of one of the two, or we must specify the ratio of the variances (Schnute 1987). However, even once we specify one of the variances or the ratio of variances, the joint estimation of observation and process uncertainty is computationally difficult and frequently ambiguous. We recommend the following:

1. Whenever possible, conduct independent experiments to determine the magnitude of observation and process uncertainties so that you will not have to estimate these from the data used in the comparison of models.
2. If possible, eliminate observation uncertainty by good experimental design or instrumentation.

3. Compare models and/or estimate parameters using the alternative, extreme assumptions of no observation uncertainty or no process uncertainty.

4. If there is little difference between your conclusions using the different assumptions in step 3, you can stop worrying about the issue. If, however, there are major differences in the results of the analysis depending on the assumption in step 3, you must either delve deeply into the statistical literature on the subject (Schnute 1987 is a good starting point) or redesign the experiments and try again.

*Likelihoods for Observation and Process Uncertainty*

While simultaneous estimation of process and observation uncertainty can be complex, the special cases in which only one is present can be analyzed in a straightforward manner.

We begin with a general deterministic model for $Y$, based on independent variables $X$ and parameters $p$,

$$Y_{\text{det}} = f(X,p) \qquad (7.17)$$

where $f(X,p)$ is assumed to be known. Now assume that the observed value of $Y$ depends on the deterministic value and the process uncertainty $W$, so that

$$Y_{\text{obs}} = Y_{\text{det}} + W. \qquad (7.18)$$

The deviation $D$ between the observed and predicted (deterministic) values of the dependent variable is

$$D = Y_{\text{obs}} - Y_{\text{det}} = W. \qquad (7.19)$$

Thus, the probability distribution of the deviation is exactly the same as the probability distribution $W$. For example, if $W$ is normally distributed, the negative log-likelihood (using $t$ as a subscript for individual observations of $X$ and $Y$) is

$$\mathbf{L}_t = \log(\sigma) + \frac{1}{2}\log(2\pi) + \frac{[Y_{\text{obs},t} - f(X_t,p)]^2}{2\sigma^2}. \qquad (7.20)$$

143

Now assume that $X$ is measured imprecisely but that $Y$ is measured exactly. In that case, the statistically interesting questions involve the value of $X$, which is related to $Y$ through the inverse function

$$X = f^{-1}(Y,p). \qquad (7.21)$$

For example, if $Y = pX$, then

$$f^{-1}(Y,p) = \frac{Y}{p}. \qquad (7.22)$$

That is, the inverse function involves "solving for $x$ in terms of $y$." This cannot always be done explicitly, and in some cases—involving nonlinear functions—the inverse function may not exist at all.

The observed value of $X$ is then

$$X_{\text{obs}} = f^{-1}(Y,p) + V, \qquad (7.23)$$

where $V$ is the observation uncertainty. Given $Y$, we calculate the predicted value of $X$ (remember the model is deterministic), and the difference between the observed $X$ and the predicted value from the inverse model is the value of the observation uncertainty. For example, if $V$ were normally distributed with mean 0 and variance $\sigma^2$, the negative log-likelihood would be

$$\mathbf{L}_t = \log(\sigma) + \frac{1}{2}\log(2\pi) + \frac{[X_{\text{obs},t} - f^{-1}(Y_t,p)]^2}{2\sigma^2}. \qquad (7.24)$$

Linear regression models are a special case of this analysis in which there is a straightforward inverse model. For a linear regression,

$$Y_{\text{det}} = f(X,p) = p_1 + p_2 X, \qquad (7.25)$$

the inverse model is

$$f^{-1}(Y,p) = \frac{Y - p_1}{p_2}. \qquad (7.26)$$

An ecological example of the linear regression Equation 7.25 is the simple model of population dynamics with survival ($s$) and births ($b$), process ($W_t$), and observation uncertainties ($V_t$) that are normally distributed with mean 0 and variance $\sigma_w$ or $\sigma_v$, respectively:

$$N_{t+1} = sN_t + b + W_t,$$

$$N_{obs,t} = N_t + V_t. \qquad (7.27)$$

When there is only process uncertainty, $N_t$ is measured perfectly and the only stochastic element affects $N_{t+1}$. The negative log-likelihood is

$$\mathbf{L}_t = \log(\sigma_W)$$
$$+ \frac{1}{2}\log(2\pi) + \frac{(N_{t+1} - b - sN_t)^2}{2\sigma_W^2}. \qquad (7.28)$$

On the other hand, if we assume only observation uncertainty, we use the inverse function method to write

$$N_{obs,t} = -\frac{b}{s} + \frac{1}{s}N_{t+1} + V_t, \qquad (7.29)$$

and the negative log-likelihood is

$$\mathbf{L}_t = \log(\sigma_V) + \frac{1}{2}\log(2\pi)$$
$$+ \frac{[N_{obs,t} + b/s - (1/s)\,N_{obs,t+1}]^2}{2\sigma_V^2}. \qquad (7.30)$$

The likelihoods in Equations 7.28 and 7.30 refer to only a single time period. The natural next question is: What should be done when time periods are linked?

## Considerations for Dynamic Models

The ecological detective often deals with observations that are a time series about the state of the system and perturbations to the system. Such time series commonly arise in wildlife, fisheries, and forestry. When the data are a time series, the model must perforce be a dynamic one in which the state of the system at a given time is linked with its values at previous times. In this section, we shall illustrate the special considerations that arise in such a case. To illustrate the ideas, we use the discrete logistic equation

$$N_{t+1} = N_t + rN_t \left( 1 - \frac{N_t}{K} \right). \tag{7.31}$$

In this equation, $N_t$ is the population size at time $t$, $r$ is the maximum possible per capita growth rate, and $K$ is the carrying capacity. We can include additions or removals ($C_t$) from the population to obtain

$$N_{t+1} = N_t + rN_t \left( 1 - \frac{N_t}{K} \right) - C_t. \tag{7.32}$$

Next, we must specify the nature of the observation and process uncertainty for this model. When the logistic model is used in practice, it is commonly (although certainly not universally) assumed that both observation and process uncertainties are log-normally distributed. This means, for example, that we assume that the observation is

$$N_{obs,t} = N_t V,$$

$$V = \exp \left( Z\sigma_V - \frac{\sigma_V^2}{2} \right), \tag{7.33}$$

where $Z$ is normally distributed with a mean of zero and a standard deviation of 1, and $\sigma_V$ is the standard deviation of the observation uncertainty (see Equation 3.68 ff. to justify the formulation).

Process uncertainty is included in a similar manner:

$$N_{t+1} = W_t \left[ N_t + rN_t \left( 1 - \frac{N_t}{K} \right) - C_t \right],$$

$$W_t = \exp \left( Z\sigma_W - \frac{\sigma_W^2}{2} \right). \tag{7.34}$$

A scenario described by this model might be the discovery of a previously unfished resource, its overexploitation, and subsequent reductions in catch to correct the problem. To describe this situation, we could use the Monte Carlo method to generate data in ten time periods (starting with an unperturbed population), allow harvesting of half of the population at times 3, 4, and 5 (the overexploitation), and reduce the harvest rate to almost zero for the last four time periods (the "management action").

Assuming the parameters $r = 0.5$, $K = 1000$, $\sigma_W = 0.1$, and $\sigma_V = 0.1$, a pseudocode is:

---

Pseudocode 7.1

1. Input values of the parameters $r$, $K$, $\sigma_W$, and $\sigma_V$.
2. Set the initial value of population at $K$.
3. Calculate population size next year based on the logistic equation with process uncertainty, harvesting half of the population at times 3, 4, and 5.
4. Calculate the observed population at each time period.
5. Repeat steps 3 and 4 for ten years.

---

The Monte Carlo method provides a trajectory of population size over time (Figure 7.4). Assuming only observation uncertainty means that we should use Equation 7.33. The deviation between the observed and true values of the logarithm of population size is

$$D_t = \log(N_{\text{obs},t}) - \log(N_t) + \frac{\sigma_V^2}{2},$$
(7.35)

and using Equation 7.33,

$$D_t = [\log(N_t) + \log(V)] - \log(N_t) + \frac{\sigma_V^2}{2}$$

$$= \log(V) + \frac{\sigma_V^2}{2}$$

$$= \left( Z\sigma_V - \frac{\sigma_V^2}{2} \right) + \frac{\sigma_V^2}{2} = Z\sigma_V.$$
(7.36)

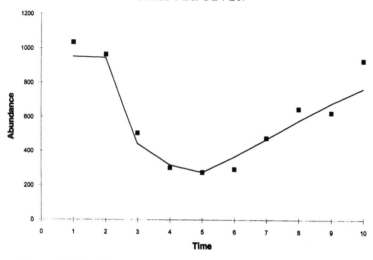

FIGURE 7.4. The Monte Carlo data (squares) for the logistic model Equation 7.34. The harvest rate is 50% during periods 3, 4, and 5, and 0.01 at other times. The line shows the best fit of the model assuming observation uncertainty. The estimated parameters are $r = 0.47$ and $K = 960$.

Thus, $D_t$ is normally distributed with mean 0 and variance $\sigma_V^2$, so that the likelihood of a deviation of size $d_t$ is

$$\mathscr{L} = \frac{1}{\sqrt{2\pi\sigma_V^2}} \exp\left( -\frac{d_t^2}{2\sigma_V^2} \right), \tag{7.37}$$

and the negative log-likelihood for the observation at time $t$ is

$$\mathbf{L}_t = \log(\sigma_V) + \frac{1}{2}\log(2\pi) + \frac{d_t^2}{2\sigma_V^2}. \tag{7.38}$$

This is analogous to Equation 7.28. The negative log-likelihood for all of the data (across multiple periods) is the sum of the $\mathbf{L}_t$ from Equation 7.38

Given the data and particular values of $r$, $K$, and $\sigma_V$, we can evaluate the likelihood of that set of parameters. Alternatively, we can select the parameters that make the negative log-likelihood as small as possible and call these the "best-fit" parameters. A pseudocode to do these calculations is:

---

Pseudocode 7.2

1. Input data values for observed population size.
2. For specified values of $r$ and $K$, systematically search over individual $r$ and $K$ values and generate predicted deterministic population sizes using Equation 7.32.
3. Calculate the deviation at each time period using Equation 7.36.
4. Calculate the negative log-likelihood of the deviations using Equation 7.38.
5. Sum the $L_t$ over $t$ to obtain the negative log-likelihood for the combination of $r$ and $K$ in question.
6. See which values of $r$ and $K$ lead to the smallest total likelihood.

---

By implementing this pseudocode, we predict a deterministic trajectory of the population conditioned on the parameters of the model and the starting population size (Figure 7.4). We assumed that the population is initially at carrying capacity; if one does not know that $N_0 = K$, the starting population size must also be estimated.

If there is only process uncertainty, the dynamic model becomes

$$N_{obs,t} = N_t,$$

$$N_{t+1} = W_t\{N_{obs,t} + rN_{obs,t}[1 - (N_{obs,t}/K)] - C_t\},$$

$$W_t = \exp\left( Z\sigma_W - \frac{\sigma_W^2}{2} \right). \tag{7.39}$$

The deviation is defined in a similar manner:

$$D_t = \log(N_{t+1}) - \log(N_{obs,t}) + \frac{\sigma_W^2}{2}$$

$$= \log(W_t) - \frac{\sigma_W^2}{2} = Z\sigma_W. \tag{7.40}$$

149

The key difference between the deviations in Equations 7.36 and 7.40 is that in Equation 7.40 the predicted value depends on the *observed* value in the previous time period, rather than on the *predicted* value in the previous time period. The negative log-likelihood for a single period is analogous to Equation 7.38:

$$\mathbf{L}_t = \log(\sigma_W) + \frac{1}{2}\log(2\pi) + \frac{d_t^2}{2\sigma_W^2}. \tag{7.41}$$

Once again, we can find the values of $r$ and $K$ that give the best fit to the data. To do so, we need a different pseudocode:

---

Pseudocode 7.3

1. Input the data values for observed population size.
2. For specified values of $r$ and $K$, generate predicted population sizes using Equation 7.39.
3. Calculate the deviation at each time period using Equation 7.40.
4. Calculate the negative log-likelihood of deviations using Equation 7.41.
5. Sum $L_t$ across $t$ to obtain the negative log-likelihood for the combination of $r$ and $K$ in question.
6. See which values of $r$ and $K$ lead to the smallest total likelihood.

---

The results (Figure 7.5) show that assuming only one kind of uncertainty provides a reasonably good fit to the data, although clearly neither of these models is "correct." This is gratifying, since the two assumptions that we considered are the "extremes" that bracket the true situation. As a general rule, when the data are informative, the assumption about how uncertainty enters does not matter greatly. In practice, the assumptions of only observation uncertainty or only process uncertainty have specific strengths and weaknesses. For instance, in order to use the assumption of pro-

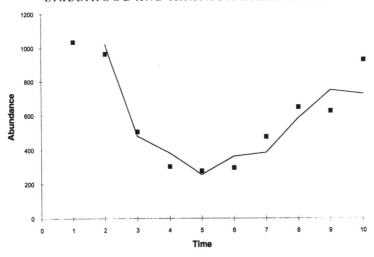

FIGURE 7.5. The same Monte Carlo data as in Figure 7.4 and the fit of the model assuming process uncertainty. The estimated parameters are $r =$ 0.44 and $K = 1023$.

cess uncertainty, we should observe each state variable at each occasion; otherwise the computation of the predicted value at future times becomes much more complex. In contrast, the assumption of only observation uncertainty makes no specific requirements about how much of the state can be observed, nor how often it is observed. The likelihood can be calculated from a single observation at any time. The major limitation of the observation uncertainty assumption is the need to specify the starting state. For example, above we assumed that $N_0 = K$. If we did not have this additional information, we would have to estimate an additional parameter, $N_0$.

The importance of the starting condition is accentuated when the model exhibits chaotic behavior, since the time trajectory of a chaotic model is highly sensitive to the starting conditions. In practice (Adkison 1992), estimators based on observation uncertainty cannot be used in chaotic models. Many models, including the discrete logistic, can

151

exhibit chaotic behavior over some range of parameters, implying that particular care is needed in formulation. Estimators based on observation tend to have trouble when dealing with long, complex time series of data. Since the observation estimator is deterministically predicted from initial conditions, if the time series has numerous changes due to random processes, observation-fitting procedures are often unable to capture the essence of the dynamics, and thus may provide poor estimates.

An additional problem for the ecological detective who works with time series is the lack of independence of the observations. Unlike true experimental situations in which the experimenter controls the state of the system, when working with time series the most we can hope for are informative perturbations. The data from one time to the next are not independent, and biases in parameters may be introduced. In practice, it is rarely possible to calculate a bias correction factor, and we recommend the use of Monte Carlo simulations to explore the sensitivity of results to the time series bias. Such simulations can be accomplished by taking the parameters estimated from the data, using them as "true" values in a Monte Carlo model, generating a few hundred data sets, and then seeing how accurately one can estimate the "true" parameters.

## MODEL SELECTION USING LIKELIHOODS

We are now ready to consider the resolution of the contest between different models for the same phenomenon, arbitrated by the data, using likelihood as the criterion. Imagine a number of models $M_1, M_2, \ldots$, in which model $M_i$ has parameters $p_{i1}, p_{i2}, \ldots$, and that we have determined the best-fit values of the parameters. In most situations, a model will rarely win the contest outright, but rather each additional experiment or observation changes our relative belief in competing models. The treatment of relative belief is covered in Chapter 9 on Bayesian methods. However, in many applications, and for many scientific journals, we must decide a winner in the contest; that is, we must choose which model appears to be "best" given the available data.

In the discussion that follows, we shall use the words "model" and "hypothesis" interchangeably. The first principle we use is that of likelihood, which quantifies how consistent a particular hypothesis is with the observations. As a general rule, the best model is the one that has the highest likelihood. When we have many competing hypotheses with the same number, of parameters, the hypothesis with the highest likelihood is the "best" one. For example, in a regression model, different slopes and intercepts are competing hypotheses, and the slope and intercept that have the highest likelihood are the best estimates of the true slope and intercept. An interesting evolutionary application of model selection using likelihood is the work of Sanderson and Donoghue (1994) in a study of the origin of angiosperms.

### The Likelihood Ratio Test for Nested Models

Commonly, the competing models do not have the same number of parameters, and a model with more parameters has an intrinsic advantage in being able to fit data. How do we referee a contest between unequal competitors, for example, between a model with one parameter and a model with two parameters? Here we rely on a second principle, known as the likelihood ratio test. The likelihood ratio test is based on the following result from theoretical statistics (Kendall and Stewart 1979, 240 ff.). Imagine two nested models, A and B, in which B is the more complicated model. That is, model B has more parameters and collapses to model A when some of them are set equal to 0. Denote the data by $Y$ and the negative log-likelihoods of the data, given the models, by $\mathbf{L}\{Y|M_A\}$ and $\mathbf{L}\{Y|M_B\}$. We assume that the more complicated model fits the data better, so that $L\{Y|M_A\} > L\{Y|M_B\}$.

The result of statistical theory is that

$$\mathfrak{R} = 2[L(Y|M_A) - L(Y|M_B)] \qquad (7.42)$$

has a chi-square distribution (refer to Chapter 3), with the degrees of freedom equal to the difference in the number of parameters between models B and A. Because the right-hand side of Equation 7.42 involves log-likelihoods, $\mathfrak{R}$ is the ratio of the logarithm of the likelihoods, and this procedure is called the likelihood ratio test.

It is perhaps easiest to understand how Equation 7.42 is used for the case of comparing the likelihood associated with a maximum likelihood estimate (MLE) parameter with the likelihood for other values of the parameters. We replace $L\{Y|M_B\}$ with $L\{Y|p_{MLE}\}$ and $L\{Y|M_A\}$ with $L\{Y|p\}$, where $p$ is another value of the parameter. The difference $\mathfrak{R}(p)$ now has a chi-square distribution with one degree of freedom, because we have one fitted parameter. If we plot the probability that $\mathfrak{R}(p)$ is less than $z$ as $p$ varies, we obtain a function that is symmetric around $p_{MLE}$ and which is zero when $p = p_{MLE}$ (Figure 7.6). The thin parabolic line is the difference in log-likelihood between $p_{MLE}$ and $p$. The thick funnel-shaped line is the probability that the $\chi^2$ random variable is less than $z = (p - p_{MLE})^2$. This plot rises to 1 as the difference between $p$ and $p_{MLE}$ increases. We construct confidence intervals by noting that $\Pr\{\chi^2 < 3.84\} = 0.95$. Consequently, if model B has one more parameter than model A, twice the difference in negative log-likelihoods must be greater than 3.84 for model B to be significantly better at the 0.05 level. We construct the confidence intervals by drawing a horizontal line at the desired confidence level (e.g., 95%) and seeing where the line intersects the $\chi^2$ probability curve. In the case of Figure 7.6, we see that this occurs at $p$ values of roughly 1 and 9. The likelihood ratio test allows us to examine models of increasing complexity to

154

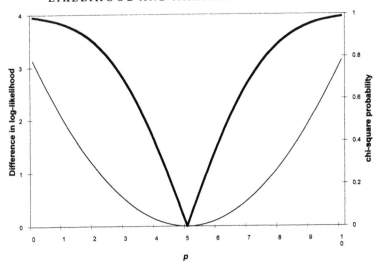

FIGURE 7.6. The relationship between the negative log-likelihood and the $\chi^2$ value used in the likelihood ratio test. The thin line is the difference in negative log-likelihoods between the best-fit parameter ($p_{MLE} = 5$) and other values of the parameter. The thick line is the probability that the $\chi^2$ random variable is less than the deviation $p$.

determine if the more complex model provides a significantly better fit.

*An Ecological Scenario.* To illustrate the use of likelihood for model selection, consider a model (Schnute 1987) relating the number of animals recorded by observers in a survey (an index of abundance $I$) to the true abundance $D$ by

$$I = \max\left\{0, \frac{p + qD}{1 + rD}\right\},\qquad (7.43)$$

where $p$, $q$, and $r$ are parameters. We obtain a series of nested models by setting one or all of the parameters equal to 0. In the simplest case, when $r = p = 0$, the index is proportional to the number of animals present with constant of proportionality $q$,

$$I = qD.\qquad (7.44)$$

155

The parameter $p$ allows for the possibility that we may conclude that even when no animals are present some are recorded ($p > 0$), or that we will not see any animals when they are rare ($p < 0$). The parameter $r$ allows for nonlinearity between the index and the true abundance. Suppose that the number of individuals observed, $I_{obs}$, is the true number plus an observation uncertainty $V$ that is Poisson distributed. Thus, $I_{obs} = I + V$ will always equal or exceed the true number because $V \geq 0$. As before, we begin by using Monte Carlo simulation to generate data in which we know the true situation:

---

Pseudocode 7.4

1. Read in values of $q = 1.0$, $r = 0.03$, and $p = -3$.
2. Set $D = 1$.
3. Calculate the deterministic values from Equation 7.43.
4. Calculate the actual observation by adding a Poisson distributed term to the result from step 3.
5. Increment the value of $D$ by 1 and repeat steps 3 and 4 until $D > 20$.

---

The squares in Figure 7.7 are the data that result from this pseudocode (Table 7.1). The dashed line is the true relationship between the index and abundance. As is typical of Poisson processes (in which the variance is equal to the mean), there is more variability at higher expected values of the index. There are four possible models:

*Model A*: Only $q$ determines the relationship (i.e., $p$ and $r$ are assumed to be equal to zero) between $D$ and $I$

*Model B*: The parameters $q$ and $p$ determine the relationship between $D$ and $I$

*Model C*: The parameters $q$ and $r$ determine the relationship between $D$ and $I$

*Model D*: All three parameters determine the relationship between $D$ and $I$

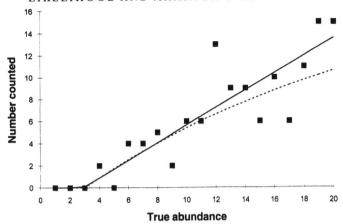

FIGURE 7.7. One set of data (squares) generated from Pseudocode 7.4 with $q = 1$, $r = 0.03$, and $p = -3$. The dashed line shows the true relationship and the solid line shows the linear model fit to the data.

TABLE 7.1.   Data generated from Pseudocode 7.4.

| Density | Index from Equation 7.43 | Number observed |
|---|---|---|
| 1 | 0 | 0 |
| 2 | 0 | 0 |
| 3 | 0 | 0 |
| 4 | 0.89 | 2 |
| 5 | 1.74 | 0 |
| 6 | 2.54 | 4 |
| 7 | 3.31 | 4 |
| 8 | 4.03 | 5 |
| 9 | 4.72 | 2 |
| 10 | 5.38 | 6 |
| 11 | 6.02 | 6 |
| 12 | 6.62 | 13 |
| 13 | 7.19 | 9 |
| 14 | 7.75 | 9 |
| 15 | 8.28 | 6 |
| 16 | 8.78 | 10 |
| 17 | 9.27 | 6 |
| 18 | 9.74 | 11 |
| 19 | 10.19 | 15 |
| 20 | 10.63 | 15 |

TABLE 7.2. Parameters and negative log-likelihoods for the four models of abundance.

| Model | Value of: q | p | r | Number of parameters | Negative log-likelihood |
|---|---|---|---|---|---|
| A | 0.586 | — | — | 1 | 42.47 |
| B | 0.793 | −2.29 | — | 2 | 38.38 |
| C | 0.393 | — | −0.023 | 2 | 40.92 |
| D | 0.987 | −2.96 | 0.0157 | 3 | 38.22 |

Given a set of data generated by the previous pseudocode, we can estimate the likelihoods for each of the four models using the following pseudocode:

---

Pseudocode 7.5

1. Input the data as in Table 7.1 and starting values for the parameters $q$, $p$, and $r$.
2. Specify which model is to be used to make predictions.
3. Compute the likelihood as follows:
   a. Cycle from $D = 1$ to 20.
   b. Calculate the predicted abundance $I_{pre}$ using Equation 7.43.
   c. Calculate the negative log-likelihood of observing $I_{obs}$ given $I_{pre}$ and add this negative log-likelihood to the total negative log-likelihood.
   d. Repeat steps a–c for each data point.
4. Sum negative log-liklihood for each data point.
5. Repeat steps 2–4 for each model.

---

We then combine the likelihood calculation with a non-linear-function minimization routine to calculate the best estimates for each model (Table 7.2). Model B reduces the negative log-likelihood by over four units by adding one parameter. Since twice the difference in likelihoods must be at least 3.84 for the models to differ at the 0.05 level, the difference in the log-likelihoods between model A and model B is clearly significant. Models B and C have the same number of parameters, so model B is clearly preferred. Model D

TABLE 7.3.   Number of times in one hundred Monte Carlo trials that each of the four abundance models was selected.

| Model | Parameters | Number of times selected with one hundred Monte Carlo data sets |
|-------|------------|----------------------------------------------------------------|
| A | $q$ only | 14 |
| B | $q$ and $p$ | 79 |
| C | $q$ and $r$ | 0 |
| D | $q$, $p$, and $r$ | 7 |

fits the data better than model B, but the difference in negative log-likelihood is very small, and not significant according to the likelihood ratio test. Therefore we conclude that for this set of data model B is the "best."

In this particular case (for the data shown in Figure 7.7), the estimation procedure failed to detect the nonlinearity between the index abundance and real abundance but did detect the non-zero intercept. When we repeat this procedure with many different Monte Carlo–generated sets of data, we find quite frequently that model A is preferred (Table 7.3).

*Akaike Information Criterion (AIC) for Non-nested Models*

The likelihood ratio test provides a simple and powerful format for comparing alternative models, but requires that the models being compared be nested, that is, the more complex model reduces to the simpler model by setting parameters equal to 0. When dealing with non-nested models, the Akaike information criterion (AIC) is normally used (Akaike 1973; Sakamota et al. 1986). Whereas the likelihood ratio test is based on an inferential criterion, the AIC is based on an optimization criterion (Akaike 1985, 1992; de-Leeuw 1992).

The AIC for model $M_i$ with $p_i$ parameters is

$$A_i = \mathbf{L}(Y|M_i) + 2p_i. \tag{7.45}$$

159

The model selection criterion is that the best model is the one that has the lowest AIC. By adding 2 to the negative log-likelihood for every free parameter, we are "penalizing" the goodness of fit in a way that is similar to the likelihood ratio test. We compare models by looking at differences in the AIC and are once again implicitly using a form of the likelihood ratio test, although the AIC is considered valid when using non-nested models.

Sakamoto et al. (1986) describe an alternative to the AIC called the Bayesian information criterion or BIC (Schwarz 1978). Hongzhi (1989) proposed an analog of the AIC for use with the sum of squares. The proposal is to use $\log(SSQ_k)$ + $2k/n$ as the analog of Equation 7.45, where $SSQ_k$ is the residual sum of squares for the model with $k$ parameters, and $n$ is the number of points. Anderson et al. (1994) evaluate the performance of the AIC for model selection in capture-recapture data. Matsumiya (1990), Hiramatsu and Kitada (1991), and Hiyama and Kitahara (1993) provide examples of the use of the AIC in fisheries problems.

## Which Criterion to Use?

The AIC is appropriate for non-nested models but for nested models either the likelihood ratio or the AIC can be used. As a note of caution, when using the Poisson or multinomial likelihoods and if the data are overdispersed, the likelihood ratio test or the AIC will be biased, and the analysis of deviance (McCullagh and Nelder 1989) is appropriate.

## ROBUSTNESS: DON'T LET OUTLIERS RUIN YOUR LIFE

Our colleague David Fournier once said, "The problem with likelihood is that some observations are just too unlikely." That is, some outliers will dominate the likelihood, and the fitting procedures often go to great lengths to make predictions closer to the outlier so that the total likelihood will not be too low.

"Robust estimation" has two meanings (Huber 1981). First, what happens if the assumption of normally distributed uncertainty is not appropriate, which is often the case for ecological data sets? Second, how does one deal with one or two data points that are highly irregular (greatly deviate from the pattern suggested by the other data)? We already discussed one approach when we considered the goodness of fit provided by the sum of squares. In that case, we noted that the use of the square of the deviation between the observed and predicted data points is implicitly based on an assumption of normally distributed uncertainty, but that other measures of deviation such as absolute value (or even fractional powers of the absolute value) could be used just as easily. Most of these have the effect of reducing the penalty which the outliers contribute to the sum of the deviations.

Another approach (Press et al. 1986, 539 ff.) is to weight the data points in the sum of squares or the likelihood. For example, one could use Tukey's "biweight"

$$\omega(e) = \text{weight assigned to uncertainty of size } e$$

$$= \left( 1 - \frac{e^2}{c^2} \right)^2 \quad \text{if } |e| < c,$$

$$= 0 \quad \text{if } |e| > c, \qquad (7.46)$$

where $c$ is a constant chosen by the user (Press et al. 1986, 542). (For normally distributed uncertainty, the appropriate value of $c$ is 6.0). This weighting function actually decreases as $e$ increases, and is consonant with the idea that outliers might be caused by something other than the actual ecological processes being studied. For example, to modify the simple sum of squares

$$\mathcal{S}(A_{\text{est}}, B_{\text{est}}, C_{\text{est}}) = \sum_{i=1}^{n} (Y_{\text{pre},i} - Y_{\text{obs},i})^2,$$

we use

$$\mathcal{G}(A_{est}, B_{est}, C_{est}) = \sum_{i=1}^{n} \omega(e_i) (Y_{pre,i} - Y_{obs,i})^2. \qquad (7.47)$$

where $e_i = Y_{pre,i} - Y_{obs,i}$. One way to think about outliers is that for any data point there is a probability $p_{model}$ that the point arose from the model that you are considering and a probability $1 - p_{model}$ that it arose from a process other than the one specified in the model. Then the likelihood of a particular point is really $p_{model}\mathcal{L}(data|model) + (1 - p_{model})\mathcal{L}(data|alternative processes)$. In general, we assume that $p_{model} = 1$. To use this approach, one needs to begin to specify what the alternative processes are; in effect, one must specify alternative models (Schnute 1993; Schnute and Hilborn 1993).

## BOUNDING THE ESTIMATED PARAMETER: CONFIDENCE INTERVALS

We must always be aware that the most likely parameters are almost certainly not the real parameters of the underlying process, but rather depend on the data. How do we determine reasonable bounds for the estimated parameter? In this section we explore two approaches to quantifying uncertainty about parameter values.

### Likelihood Profile

Hudson (1971) provides an especially simple method for determining a confidence bound for the case in which (i) we consider a model with only one parameter and (ii) the log-likelihood function is a unimodal function of the parameter. Hudson's method is a special case of the general technique of the likelihood profile. Using the likelihood ratio test (the theory relies, once again, on the asymptotic relationship followed by the differences in log-likelihood), the

95% confidence interval is the range of parameters for which the log-likelihood is within 1.92 of the maximum value of the log-likelihood. Thus, for example, to find the confidence interval for the Poisson rate parameter for the negative log-likelihoods shown in Figure 7.1c, we draw a horizontal line at the minimum negative log-likelihood plus 1.92 (the critical value of $c^2$ with one degree of freedom divided by 2) and look for the intersections of that line and the curve. Those intersection points give the limits of the confidence interval.

*An Ecological Scenario.* Suppose that we are involved in the control of mites that attack pistachios and have decided that if fewer than 10% of the nuts are attacked, the mite is being controlled. We want to determine the proportion infested ($f$) by sampling nuts. If the true level of infestation is $f$ and we sample $S$ nuts, the number $I$ that are infested follows a binomial distribution:

$$\Pr\{I = i|f\} = \binom{S}{i} f^i (1 - f)^{S-i}. \tag{7.48}$$

If we view this as the likelihood of values of $f$, given $S$ and $i$, the negative log-likelihood is

$$\mathbf{L}\{S,i|f\} = -i\log(f) - (S - i)\log(1 - f) + J, \tag{7.49}$$

where $J$ denotes terms that do not depend on $f$ and can therefore be ignored. Setting the derivative of $\mathbf{L}\{S,i|f\}$ with respect to $f$ equal to 0 leads us to the MLE value

$$f_{\text{MLE}} = \frac{i}{S}. \tag{7.50}$$

We use the likelihood ratio test to determine the approximate 95% confidence interval for $f$ by finding the value of $f$ such that the log-likelihood $\mathbf{L}\{S,i|f\} - \mathbf{L}\{f_{\text{MLE}}|S,i\} = 1.92$. Furthermore, we can do this with a sequential sampling scheme, as in the following pseudocode:

---

Pseudocode 7.6

1. Set $S = 0$, $i = 0$.
2. Input the number of nuts sampled and the number of sampled nuts that were infested. Replace $S$ by $S$ plus the number of sampled nuts and $i$ by $i$ plus the number of infested nuts in the current sample.
3. Find the MLE value $f_{\text{MLE}} = i/S$. Find the negative log-likelihood associated with this MLE from Equation 7.49.
4. Find the value of $f_b$ such that

$$\mathbf{L}\{f_b|S,i\} = \mathbf{L}\{f_{\text{MLE}}|S,i\} + 1.92.$$

If this value of $f \leq 0.1$, declare the mite under control. Otherwise return to step 2.

---

A typical set of results using this pseudocode would be these.

| Sample number | Current sample | Infested nuts | Total sample | Total infested | $f_{\text{MLE}}$ | $f_b$ |
|---|---|---|---|---|---|---|
| 1 | 20 | 2 | 20 | 2 | 0.1 | 0.283 |
| 2 | 20 | 1 | 40 | 3 | 0.075 | 0.186 |
| 3 | 20 | 1 | 60 | 4 | 0.067 | 0.151 |
| 4 | 20 | 0 | 80 | 4 | 0.05 | 0.114 |
| 5 | 20 | 0 | 100 | 4 | 0.04 | 0.092 |

Note that after the first sample, the MLE is already 0.1, but the boundary of the 95% confidence interval for the true value of $f$ is 0.283, so that we must continue sampling. It is only at sample 5, for which the MLE is 0.04 and the boundary of the confidence interval is 0.092, that we can declare the mite under control. Now, of course, had we sampled one-hundred nuts at the start and found four of them infested, we would draw the same conclusion as was done after the fifth sample. The advantage of the sequential sampling scheme, using likelihood, is that we might be able to stop even sooner.

The likelihood profile can be extended for situations in which the model has more than one parameter. For example, in the abundance model Equation 7.43, the best model had two free parameters, $q$ and $p$. In such a case, we might want to know about the confidence intervals for $q$ and $p$, either separately or together.

To conduct a likelihood profile for a system with parameters $p_1$, $p_2$, . . . ,$p_m$, one varies one (or more) parameter(s) systematically and computes the values of the other parameters that maximize the likelihood. It has the same function as a goodness-of-fit profile, giving information concerning how the parameters depend on each other, and how sensitive the likelihood is to the systematically varied parameter (Venzon and Moolgavkar 1988).

For example, suppose that the random variables $X_1$, . . . , $X_n$ are normally distributed with mean $m$ and standard deviation $\sigma$. The negative log-likelihood is then

$$L = n[\log(\sigma) + \frac{1}{2} \log(2\pi)] + \sum_{i=1}^{n} \frac{(X_i - m)^2}{2\sigma^2}, \tag{7.51}$$

from which we determine the maximum likelihood estimates,

$$m_{\text{MLE}} = \frac{1}{n} \sum_{i=1}^{n} X_i \quad \text{and}$$

$$\sigma^2_{\text{MLE}} = \frac{1}{n} \sum_{i=1}^{n} (X_i - m_{\text{MLE}})^2. \tag{7.52}$$

A likelihood profile is appropriate for a situation where we are interested in one parameter but not particularly interested in the other. If the parameter of interest is the mean, we systematically search over values of $m$ and instead of $\sigma_{\text{MLE}}$, we compute the profile standard deviation

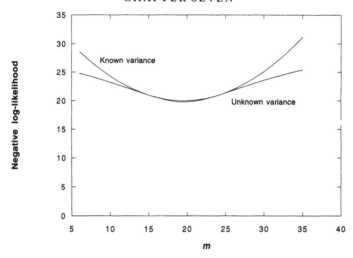

FIGURE 7.8. The negative log-likelihood for the mean $m$ of a normal distribution when the variance is known, and the profile likelihood in which the variance is specified once the mean is given. Note that both the negative log-likelihood and likelihood profile find the mean, but that the likelihood profile is shallower (more uncertainty) when the variance is unknown.

$$\sigma^2_{\text{pro}} = \frac{1}{n} \sum_{i=1}^{n} (X_i - m)^2.$$

(7.53)

For example, if the data are 27.7286, 16.4676, 21.1222, 27.6477, 10.4809, and 13.9685 (generated from $m = 20$ and $\sigma = 8$), plots of the negative log-likelihood and likelihood profile find the true mean (Figure 7.8), but admitting that the standard deviation is unknown leads to a shallower negative log-likelihood and consequently to a wider confidence interval.

*An Ecological Scenario.* To find the likelihood profile for $q$ for the abundance model Equation 7.43, we find the values of $p$ and $r$ that maximize the likelihood for each possible value of $q$ (or, in reality, a grid search over $q$), as in the following pseudocode:

166

Pseudocode 7.7

1. Input the lower and upper bounds, and the step size of $q$ to search.
2. Set $q$ fixed at the lower bound.
3. Choose one of:

   Option *a*. Calculate the negative log-likelihood by using true values of $r$ and $p$.

   Option *b*. Minimize the negative log-likelihood by searching over possible values of $r$ and $p$ (the true likelihood profile).
4. Plot or table the values of $q$ and the negative log-likelihood.
5. Increment $q$ and repeat steps 3 and 4.

This algorithm allows for two cases. First (step 3, option *a*), we fix the other parameters ($r$ and $p$) at their true values (known because we have used Monte Carlo data) and examine the likelihood in $q$. This will demonstrate how much more we would know about $q$ if the values of $r$ and $p$ were known. That is, instead of the MLE values, we use the true values of the other parameters. The results (Figure 7.9, dashed line) are quite impressive. The confidence interval for $q$ is very narrow. Second (step 3, option *b*), we find the $r$ and $p$ that maximize the likelihood as $q$ is systematically varied; this is the likelihood profile. The results (Figure 7.9, solid line) are discouraging. We can fit the data very well (i.e., the negative log-likelihood is small) with very large values of $q$. For example, the dashed line in Figure 7.10 shows the fit obtained when $q = 10$, $p = -40$, and $r = 1.58$. This curve is very similar to the true relationship, but clearly the individual parameter values are far from the true values (recall that a similar phenomenon occurred in Chapter 5). The confidence bound on $q$ is enormous. In effect, admitting uncertainty in $p$ and $r$ means that we know nothing about the value of the individual parameter $q$.

167

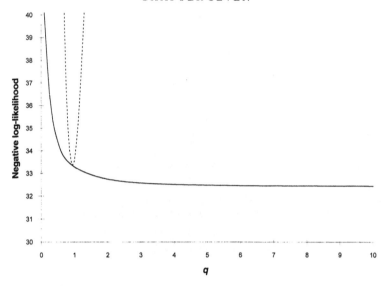

FIGURE 7.9. Likelihood profiles of $q$ when $p$ and $r$ are estimated parameters (solid line) and when $p$ and $r$ are fixed at their true values (dashed).

## THE BOOTSTRAP METHOD

In Chapters 5 and 6, we used the bootstrap method to resample data sets for model comparison. Here we extend its use for understanding the uncertainty about parameter values. The bootstrap method can be used to find confidence intervals and variances of models of any complexity by intense computation (Efron and Tibshirani 1991, 1993). As before, the bootstrap method involves generation of new data sets by sampling the original data with replacement. We begin with a set of $N$ observations $\{Y_i, \ldots, Y_N\}$. We generate a large number of new data sets $\{Y_{boot}(i)\}$ by sampling from $Y_{obs}$ with replacement and then generate a large number of bootstrap data sets. For each bootstrap data set we obtain an estimate of the parameters of interest and estimate the variances of the parameters from the variances of the bootstrap estimates.

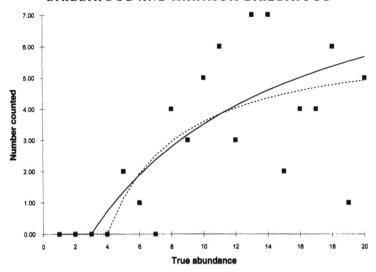

FIGURE 7.10. Data generated by the Monte Carlo method for the abundance model Equation 7.43. The true relationship is shown by the solid line, and a model with $q = 10$, $p = -40$, and $r = 1.58$ is shown by the dashed line.

Suppose that there is just one parameter, that we generate $B$ bootstrap data sets, and that $p_{boot,i}$ is the parameter estimate from the $i^{th}$ bootstrap data set. We first set

$$\bar{p}_{boot} = \sum_{i=1}^{B} p_{boot,i} / B. \tag{7.54}$$

We estimate the variance by

$$\hat{\sigma}_p = \frac{1}{B-1} \sum_{i=1}^{B} (\bar{p}_{boot} - p_{boot,i})^2. \tag{7.55}$$

Returning once again to the abundance model Equation 7.43, we might want to use the bootstrap method to estimate the variance of the parameter $q$. This can be done using a pseudocode such as:

169

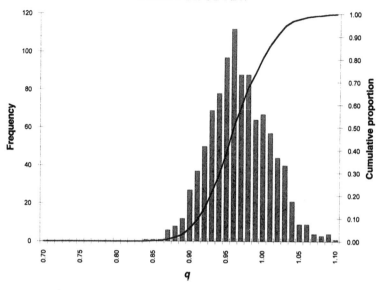

FIGURE 7.11. The distribution of estimates of $q$ from one thousand bootstrap replicates. The solid line is the cumulative distribution function.

---

Pseudocode 7.8
1. Read in observed densities and index of abundance from Table 7.1
2. Set $r = 0$, $p = 0$.
3. Generate a bootstrap data set by sampling with replacement from the data twenty pairs of $D_i$ and $I_{obs,i}$.
4. Obtain the maximum likelihood estimate of $q$ from the bootstrap data.
5. Repeat steps 3–5 1000–10 000 times.
6. Plot the frequency distribution of the estimated $q$ values.

---

The output of a program based on this algorithm is a frequency distribution of estimates of $q$ (Figure 7.11). Given a variance estimate from Equation 7.55, we can calculate the confidence bounds in the usual manner using normal distribution theory, or we can use the empirical frequency distribution of the bootstrap estimates. In the latter case, the bootstrap provides a link between the likelihood methods in this chapter and the Bayesian methods of Chapter 9.

The bootstrap method as described here is often called the non-parametric bootstrap. A refinement, based on some knowledge of the ecological system, is to assume a distribution for the uncertainty; instead of resampling the data we add a random term to the predicted data based on the assumed distribution. That is, we now generate bootstrap data sets by taking the $i^{th}$ observation $Y_{pre,i}$ and adding a random variable $E$ to it:

$$Y_{boot,i} = Y_{pre,i} + E, \tag{7.56}$$

where $E$ is drawn from the assumed distribution. In principle, this should be "better" because we are incorporating more knowledge about the system into the methods of estimation. We leave it to you to modify the previous pseudocode for the case in which $E$ has a Poisson distribution. Doing this leads to a different frequency distribution of bootstrap estimates (Figure 7.12)

Bootstrapping is a computationally intensive procedure, but it can be used on models that have dozens or even hundreds of parameters. Obtaining an estimate for large models may take minutes or even hours. It is not unknown for bootstrap runs to take several days on desktop computers, and obtaining a 99% confidence interval requires about 10 000 bootstrap samples (Efron and Tibshirani 1991, 1993).

## LINEAR REGRESSION, ANALYSIS OF VARIANCE, AND MAXIMUM LIKELIHOOD

The statistical tools learned in introductory courses in biometrics were designed in an age when computation was difficult (Efron and Tibshirani 1991), but things are different today. We now show that they can be performed using the methods of maximum likelihood and the likelihood ra-

171

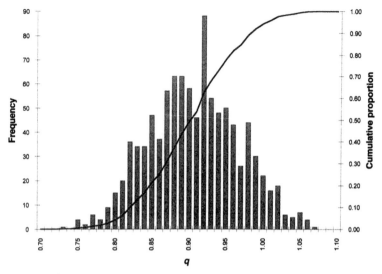

FIGURE 7.12. Estimates of $q$ from one thousand replicates of the parametric bootstrap.

tio test but in a numerically intensive manner, thus taking advantage of modern computing technologies.

It is easier to understand statistics within the unifying concept of likelihood rather than thinking of regression, analysis of variance, and contingency tables as intellectually separate subjects.

### Regression as a Problem of Maximum Likelihood

The linear regression model is

$$Y_i = a + bX_i + Z_i, \tag{7.57}$$

where the parameters $a$ and $b$ are to be determined and $Z_i$ is normally distributed with mean 0 and variance $\sigma^2$. Proceeding as before, the negative log-likelihood is

$$\mathbf{L} = n[\log(\sigma) + \frac{1}{2}\log(2\pi)]$$

$$+ \frac{1}{2\sigma^2} \sum_{i=1}^{n} (Y_i - a - bX_i)^2. \tag{7.58}$$

172

A nonlinear search over $a$, $b$, and $\sigma$ can be used to minimize the negative log-likelihood. However, the maximum likelihood estimates of $a$ and $b$ are solutions of the linear equations

$$\sum_{i=1}^{n} Y_i = n a_{\text{MLE}} + b_{\text{MLE}} \sum_{i=1}^{n} X_i,$$

$$\sum_{i=1}^{n} X_i Y_i = a_{\text{MLE}} \sum_{i=1}^{n} X_i + b_{\text{MLE}} \sum_{i=1}^{n} X_i^2, \tag{7.59}$$

found by taking the derivative of the likelihood with respect to $a$ or $b$ and setting it equal to zero.

Note that these are independent of the variance, which we estimate by

$$\sigma_{\text{MLE}}^2 = \frac{1}{n} \sum_{i=1}^{n} (Y_i - a_{\text{MLE}} - b_{\text{MLE}} X_i)^2. \tag{7.60}$$

A two-dimensional confidence interval on $a$ and $b$ is found by searching over all values of $a$ and $b$ that provide likelihoods within a specified value of the minimum negative log-likelihood. For example, for a 95% confidence interval, we use a chi-square distribution with two degrees of freedom, for which the critical value is 6.0. Thus, we contour all negative log-likelihoods that are three greater than the best value.

On the other hand, we might be interested in a single parameter, say $b$, and not at all interested in the other parameter, so that a likelihood profile on $b$ is appropriate. We first specify $b$ in the negative log-likelihood and then compute that value of $a$ that minimizes the negative log-likelihood for that value of $b$. This can be done from Equation 7.59:

$$a_{\text{pro}} = \frac{\sum_{i=1}^{n} Y_i - b \sum_{i=1}^{n} X_i}{n}. \tag{7.61}$$

Since this is now a one-parameter confidence bound, the critical chi-square value is 3.84, so values of negative log-likelihood that are 1.92 greater than the minimum are in the 95% confidence interval.

To illustrate these ideas, we generated data from the model $Y_i = 1 + 2X_i + Z_i$, with $\sigma = 5$. A typical set of ten data points is:

| $X_i$ | $Y_i$ |
|---|---|
| 1 | 7.830 37 |
| 2 | 2.792 27 |
| 3 | 7.701 37 |
| 4 | 13.779 8 |
| 5 | 5.050 55 |
| 6 | 9.230 33 |
| 7 | 3.452 11 |
| 8 | 11.952 8 |
| 9 | 23.855 9 |
| 10 | 22.088 5 |

for which $a_{MLE} = 1.77$, $b_{MLE} = 1.641$, $\sigma_{MLE} = 5.69$, and the minimum negative log-likelihood is 30.5738.

The 95% confidence contour for both parameters (Figure 7.13) is an ellipse with a negative correlation between the estimated values of $a$ and $b$. The data allow $a$ to be large, but then $b$ must be small, and vice versa. The likelihood profile on $b$ (Figure 7.14) considerably tightens the confidence region.

A good ecological detective will recognize that there are other models, such as

$$Y_i = k + Z_i \quad \text{(average value model)},$$

$$Y_i = a + bX_i + cX_i^2 + Z_i$$
$$\text{(quadratic regression model). (7.62)}$$

We encourage you to compute the negative log-likelihoods for these other models with one and three parameters, re-

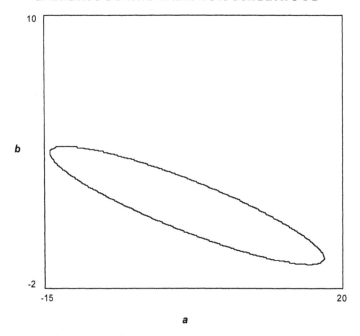

FIGURE 7.13. The 95% confidence region, determined by maximum likelihood analysis, for the parameters $a$ and $b$ of the linear regression model.

spectively, and compare the results with the regression model that we analyzed (two parameters). Which model would you choose on the basis of a likelihood criterion?

Regression methods also usually report the "proportion of variance explained by the model." Here, likelihood methods provide little additional information. However, Bayesian methods tell us that we should not attempt to "explain variation"; instead, we should construct posterior probability densities and ask about the shape of those distributions. After reading Chapter 9, we encourage you to rethink this analysis from a Bayesian perspective. What kind of priors would you choose for $a$ and $b$?

Finally, we encourage you to experiment with a situation in which we do not know the underlying model. Sokal and

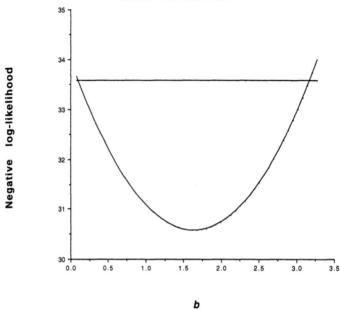

*b*

FIGURE 7.14. The likelihood profile of the parameter *b* and the 95% confidence region (below the solid line) in the linear regression model.

Rohlf (1969) report experiments in which twenty-five individual flour beetles were starved for six days at nine different humidities. The data are:

| Relative humidity (%) | Average weight loss (mg) |
| --- | --- |
| 0 | 8.98 |
| 12 | 8.14 |
| 29.5 | 6.67 |
| 43 | 6.08 |
| 53 | 5.9 |
| 62.5 | 5.83 |
| 75.5 | 4.68 |
| 85 | 4.2 |
| 93 | 3.72 |

Since the weight loss shows a clear trend with relative humidity, a linear regression model might be appropriate. What can you conclude about these data?

*Analysis of Variance by Maximum Likelihood*

TABLE 7.4. Mosquito wing lengths (Sokal and Rohlf 1969).

| Cage | Female | Left wing length measurement | |
|------|--------|-------|--------|
|      |        | First | Second |
| 1 | 1 | 58.5 | 59.5 |
| 1 | 2 | 77.8 | 80.9 |
| 1 | 3 | 84.0 | 83.6 |
| 1 | 4 | 70.1 | 68.3 |
| 2 | 5 | 69.8 | 69.8 |
| 2 | 6 | 56.0 | 54.5 |
| 2 | 7 | 50.7 | 49.3 |
| 2 | 8 | 63.8 | 65.8 |
| 3 | 9 | 56.6 | 57.5 |
| 3 | 10 | 77.8 | 79.2 |
| 3 | 11 | 69.9 | 69.2 |
| 3 | 12 | 62.1 | 64.5 |

We now show how a traditional analysis of variance can be performed using maximum likelihood theory. Sokal and Rohlf (1969) describe an experiment in which twelve field-caught mosquito pupae were reared in three different cages, four mosquitoes to each cage. When the mosquitoes hatched, the left wing of each mosquito was measured twice (Table 7.4). The observations are thus the wing length $L_{ij}$ for female $i$ on observation $j$, and the cage in which female $i$ is reared, $c_i$. We postulate three different models:

$$L_{ij} = K + Z_{ij} \quad \text{(model A)},$$

$$L_{ij} = D_{c_i} + Z_{ij} \quad \text{(model B)},$$

$$L_{ij} = F_i + Z_{ij} \quad \text{(model C)}. \quad (7.63)$$

In each model, $Z_{ij}$ is normally distributed. The alternatives are (i) the observations are normally distributed about some constant $(K)$ (model A); (ii) there is a different average length $(D_{c_i})$ within each cage (model B); or (iii) there is a different average length $(F_i)$ for each individual fly (model C).

The likelihoods for the three models are

$$\mathcal{L}_A = \prod_{i=1}^{12} \prod_{j=1}^{2} \frac{1}{\sigma_A \sqrt{2\pi}} \exp\left( - \frac{[L_{ij} - K]^2}{2\sigma_A{}^2} \right),$$

$$\mathcal{L}_B = \prod_{i=1}^{12} \prod_{j=1}^{2} \frac{1}{\sigma_B \sqrt{2\pi}} \exp\left( - \frac{[L_{ij} - D_{c_i}]^2}{2\sigma_B{}^2} \right),$$

$$\mathcal{L}_C = \prod_{i=1}^{12} \prod_{j=1}^{2} \frac{1}{\sigma_C \sqrt{2\pi}} \exp\left( - \frac{[L_{ij} - F_i]^2}{2\sigma_C{}^2} \right). \tag{7.74}$$

In principle, each model has a different standard deviation. When computing the negative log-likelihoods for the three models (Table 7.5), model A requires two parameters (the global mean and the standard deviation); the standard deviation can be obtained analytically. Model B requires four parameters, a mean for each cage, and a standard deviation. Finally, model C requires a mean for each of the twelve flies and a standard deviation.

TABLE 7.5. Analysis of variance by maximum likelihood for the mosquito data.

| Model | Number of parameters | Negative log-likelihood | Chi-square probability[a] |
|---|---|---|---|
| A (Average) | 2 | 89.32 | — |
| B (Cage effect) | 4 | 85.42 | 0.02 |
| C (Female effect) | 13 | 28.90 | ~0.0000 |

[a]Used to compare models A and B (with two degrees of freedom) and models B and C (with nine degrees of freedom).

When comparing models A and B, the negative log-likelihood is reduced by about four by adding two additional parameters. Twice the difference in the likelihood between model A and model B is 7.8. The chi-square probability of a change in 7.80 with two degrees of freedom is about 0.02, so the significance of the difference is borderline (significant at 0.05 but not at 0.01). Comparing models B and C, however, we find a considerable reduction in the negative log-likelihood and an associated chi-square probability that is essentially zero. We therefore conclude that there are differences between females and that model C is preferred.

# Conservation Biology of Wildebeest in the Serengeti

## MOTIVATION

The Serengeti ecosystem, in Tanzania and Kenya, is home to the largest migratory ungulate populations in the world, as well as many other species, some rare and endangered. This ecosystem is dominated by the wildebeest or gnu (*Conochaetes taurinus*), whose population size between 1978 and 1990 was about 1.5 million individuals (Figure 8.1). Long-term research in the Serengeti began with the Grzimeks' (1960) book *Serengeti Shall Not Die*, which led to the creation of the Serengeti Research Institute (SRI), now known as the Serengeti Wildlife Research Centre (SWRC). Sinclair and Norton-Griffiths (1979) and Sinclair and Arcese (1995) document the history of research in the Serengeti.

In this chapter we consider two questions that correspond to two periods of examination of the wildebeest population trends and specific questions considered important in those periods. First, in 1978 when the herd first exceeded 1 million individuals, there was serious concern about the population if a series of dry years should occur. Second, in the early 1990s, population size had leveled but was subject to considerable illegal harvest. Managers were interested in determining the level of harvest and the potential response of the herd to increases in such uncontrolled harvest. Answering these questions shows how likelihood methods can be used to select between different models, how different sources of data can be combined through models based on

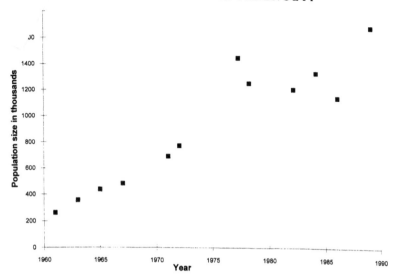

FIGURE 8.1. Abundance estimates of wildebeest population size, based on aerial surveys.

observation uncertainty and how data may be informative or not depending upon the particular question we ask.

## THE ECOLOGICAL SETTING

The wildebeest is a classic "keystone" species (Krebs 1994; but see Mills et al. 1993; models of wildebeest in the Serengeti can be found in Hilborn and Sinclair 1979). It is a large herbivore that is the major food source for the two large carnivores, lions and hyenas, and provides the bulk of the carrion for scavengers. Grazing wildebeest affect the abundance of grass, the frequency and intensity of fire, and the regeneration of trees and brush. In the last forty years, the size of the herd increased dramatically. In 1961, the estimate of wildebeest population size was 263 000; by 1977 it was 1 444 000. This increase was traced to two causes. First, in the 1950s and early 1960s rinderpest, a virus that affects ruminants and which had been introduced to east Africa through European cattle, was eliminated from cattle as a result of a vaccination program. The virus was unable to maintain itself in wild ruminants and became extinct. Second, from 1971 to 1978 there was an unusual series of years with high rainfall during the normally dry season from July to October; this increased wildebeest population size by providing additional food and associated higher survival.

In the dry season, wildebeest concentrate in the woodlands (about 1 million ha, equivalent to a square of 100 km on each side) in the north and west of the park, where there is more rainfall and more fresh grass. The plains, where the wildebeest spend the wet season and calve, are dry and barren in the dry season, and few, if any, ungulates can stay alive. When the wildebeest arrive in the woodlands there is usually a large standing stock of dry grass that grew during the wet season. The protein content of older grass is so low that a wildebeest would starve to death eating only that, but there is some rainfall during the dry season, leading to fresh growth that provides the needed protein.

## THE DATA

The principal source of data concerning wildebeest population size is a series of air surveys conducted by SRI and SWRC (Tanzania Wildlife Conservation Monitoring 1994). These surveys take place in the wet season when the entire herd is concentrated on the treeless plains and easily visible. In addition to the surveys of herd abundance, measurements of rainfall (Figure 8.2), food availability, calving rates, calf survival, and adult mortality have been collected (Table 8.1).

One of the key studies before 1978 was the relationship between dry season rainfall and fresh growth of grass (Sinclair 1979). Grass growth ($G$, measured in kg/ha·mo) during the dry season was proportional to rainfall ($R$, measured in mm)

$$G = 1.25R. \tag{8.1}$$

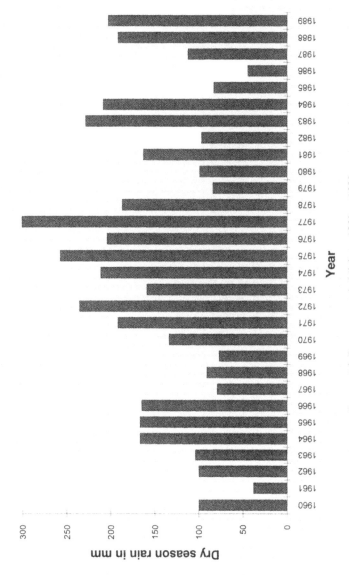

FIGURE 8.2. Dry season rainfall from 1960 to 1990.

TABLE 8.1. Serengeti wildebeest data.

| Year | Dry season rainfall (mm) | Estimated wildebeest population size (thousands) | Standard deviation of wildebeest population estimate | Estimated adult monthly dry season mortality | Estimated calf survival |
|------|------|------|------|------|------|
| 1960 | 100 | | | | |
| 1961 | 38 | 263 | | | |
| 1962 | 100 | | | | |
| 1963 | 104 | 357 | | | |
| 1964 | 167 | | | | 0.50 |
| 1965 | 167 | 439 | | | 0.25 |
| 1966 | 165 | | | | 0.30 |
| 1967 | 79 | 483 | | | 0.26 |
| 1968 | 91 | | | 0.017 | 0.36 |
| 1969 | 77 | | | 0.014 | |
| 1970 | 134 | | | | 0.32 |
| 1971 | 192 | 693 | 28.8 | 0.008 | 0.35 |
| 1972 | 235 | 773 | 76.7 | 0.005 | 0.36 |

| Year | | | | |
|------|-----|------|-----|-------|
| 1973 | 159 | | | |
| 1974 | 211 | | | |
| 1975 | 257 | | | |
| 1976 | 204 | | | |
| 1977 | 300 | 1444 | 200 | 0.027 |
| 1978 | 187 | 1249 | 355 | 0.021 |
| 1979 | 84  | | | |
| 1980 | 99  | | | |
| 1981 | 163 | | | |
| 1982 | 97  | 1209 | 272 | |
| 1983 | 228 | | | |
| 1984 | 208 | 1338 | 138 | |
| 1985 | 83  | | | |
| 1986 | 44  | 1146 | 133 | |
| 1987 | 112 | | | |
| 1988 | 191 | | | |
| 1989 | 202 | 1686 | 176 | |

During the dry season almost all rain falls in patchy thundershowers. A few days after a shower, a location turns green and the wildebeest (and other ungulates) move into the area. In addition, there are river valleys that provide some moisture and fresh grass. In the models, we assume homogenous grass production, but this is a simplification of a spatially complex system. From the rainfall-grass relationship and a total area of the woodlands, we estimate total green grass production, and by dividing by the number of wildebeest we can estimate the amount of green grass per animal.

Sinclair (1979) estimated calf survival by measuring cow-to-calf ratios at different times of the year. The key feature of the calf survival data is that over the range of observed food availability there seems to be no relationship between calf survival and food. Sinclair (1979, and personal communication) also measured monthly adult mortality rates during the dry season using transects to determine number alive and number dying per day. These data are available for 1968, 1969, 1971, 1972, 1982, and 1983. In addition, we know that in 1978 the birth rate was approximately 0.4 per wildebeest one year or older.

Thus the basic population data are occasional population censuses, calf survival for eight years, and adult mortality rate estimates for six years. The censuses conducted in 1971, 1972, 1977, and 1978 included estimates of the variance, whereas the census methods used in the 1960s did not have variance estimates. For the censuses in the 1970s, we used the published standard deviations. For the censuses in the 1960s we set the CV equal to 0.3, which is about twice the average CV from the censuses in the 1970s and 1980s and reflects a lower confidence in the early censuses (Sinclair, personal communication). The standard deviations for the censuses in the 1970s were derived from asymptotic normal theory, and in keeping with this we assume that the census data are normally distributed. We recognize, however, that this is a convenient assumption rather than demonstrated to be the case. Finally, we have rainfall data and a relationship between rainfall and dry season grass production.

It is even more difficult to determine the appropriate distribution and variance for the calf survival and adult mortality estimates. Since a survival rate can be viewed as the product of many individual survival rates over shorter periods of time (Hilborn and Walters 1992, 264 ff.), we assume that each mortality rate is log-normally distributed with CV about 0.3.

## THE MODELS: WHAT HAPPENS WHEN RAINFALL RETURNS TO NORMAL (THE 1978 QUESTION)?

We separate the models and confrontations according to the two questions described in the introduction.

Because the herd increased quite rapidly in the 1960s and 1970s, and the 1970s had been unusually wet, there was great concern in 1978 that if rainfall returned to normal (150 mm/year rather than 250 mm/year), a large portion of the herd would die. Stated in terms of a battle between hypotheses, the competing hypotheses are that (i) the herd will collapse if dry season rainfall is 150 mm for several years after 1978 and (ii) the herd will not collapse.

### A Logistic Model

We begin with a deterministic logistic model

$$N_{t+1} = N_t + rN_t \left( 1 - \frac{N_t}{K} \right),$$
(8.2)

where the number of individuals $N_t$ is measured in thousands.

Since the estimates of abundance were sporadic, it is much more difficult to use a model with process error. To

187

see this, remember that if we wanted to use a model with process error, we would multiply the right-hand side of Equation 8.2 by $\exp(W_t - \sigma^2/2)$, where $W_t$ is normally distributed with mean 0 and variance $\sigma^2$. If we missed just one observation, then instead of Equation 8.2, we require an equation relating $N_t$ and $N_{t+2}$. With many years between observations, the distributions of the sums of process uncertainties become unwieldy. They can be treated with approximate methods (a more advanced topic) or with Monte Carlo simulation.

Thus, we use a model with observation error and assume that the starting biomass in 1961 was the survey estimate of 263 000. Consequently, to Equation 8.2 we append the observation model

$$N_{\text{obs},t} = N_t + V_t, \tag{8.3}$$

Where $V_t$ is normally distributed with mean 0 and standard deviation $\sigma_t$. The negative log-likelihood in a single period is

$$\mathbf{L}_t = \log(\sigma_t) + \frac{1}{2}\log(2\pi) + \frac{(N_{\text{obs},t} - N_t)^2}{2\sigma_t^2}. \tag{8.4}$$

With this model, the only usable data are the censuses, for which there is a different $\sigma_t$ in each year when a census was conducted.

### A Life History Model

In 1978 the concern was that the most likely victims of low-rainfall years would be calves. The logistic model, with a focus on total population size, cannot capture this concern. Rather, we need more biological detail. One such life history model begins with

$T_t$ = total food produced (kg/ha·month) in the dry season
$\phantom{T_t}$ = $1.25R_t$, \hfill (8.5)

where $R_t$ is total dry season rainfall (mm). The food per animal, $F_t$, (kg/animal·month) is related to food production per ha, the total area $A$ used in the dry season (1 000 000 ha), and the number of animals $N_t$ at the start of year $t$ by

$$F_t = \frac{T_t A}{N_t}.$$ 
$$(8.6)$$

The number of births $B_t$ in year $t$ is

$$B_t = 0.4 N_t,$$ 
$$(8.7)$$

and survival of calves, $s_{\text{calf},t}$, from birth in year $t$ to their first birthday is assumed to be

$$s_{\text{calf},t} = \frac{a F_t}{b + F_t}.$$ 
$$(8.8)$$

In this equation, the parameters $a$ and $b$ determine how calf survival is related to food. In particular, $a \leq 1$ is the maximum value of calf survival and $b$ is the value of food per individual at which survival is 50% of $a$. Equation 8.8 is a Holling type-II functional response (Krebs 1994), in which the amount of food ingested is a saturating function of the amount of food available (Figure 8.3)

We use a similar functional form for the relationship between adult survival $s_{\text{adult},t}$ and food availability:

$$s_{\text{adult},t} = \frac{g F_t}{f + F_t},$$ 
$$(8.9)$$

where $g$ and $f$ have similar interpretations.

Combining all of these, we arrive at the model for population dynamics and observation:

$$N_{t+1} = (s_{\text{adult},t}) N_t + (s_{\text{calf},t}) B_t,$$

$$N_{\text{obs},t} = N_t + V_t.$$ 
$$(8.10)$$

Adult mortality in year $t$ is $M_{\text{adult},t} = (1 - s_{\text{adult},t}) N_t$.

The likelihood has three components, derived from the census, the calf survival, and the adult mortality data, re-

189

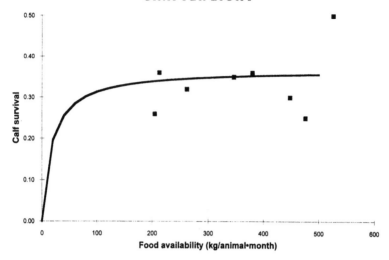

FIGURE 8.3. The relationship between calf survival and food per animal. The solid line is the best-fit Holling type-II functional response.

spectively. Because we are not estimating any of the variances from the data, we can ignore the normalization constant in the negative log-likelihood and write

$$\mathbf{L}_{\text{total}} = \mathbf{L}_{\text{census}} + \mathbf{L}_{\text{calf survival}} \\ + \mathbf{L}_{\text{adult mortality}},$$

$$\mathbf{L}_{\text{census}} = \sum_t \frac{(N_t - N_{\text{obs},t})^2}{2\sigma_1^2},$$

$$\mathbf{L}_{\text{calf survival}} = \sum_t \frac{(s_{\text{calf},t} - s_{\text{calf,obs},t})^2}{2\sigma_2^2},$$

$$\mathbf{L}_{\text{adult mortality}} = \sum_t \frac{(M_{\text{adult},t} - M_{\text{adult,obs},t})^2}{2\sigma_3^2}, \tag{8.11}$$

where $\sigma_1$, $\sigma_2$, and $\sigma_3$ are the standard deviations associated with the census, calf survival, and adult mortality, respectively; they may also depend on time.

190

Let us explore this model in some detail before involving the data. The model allows for both calf survival and adult survival to decrease as food per animal decreases. The key parameters are $a$, $b$, $g$, and $f$, and the population dynamics model can be written as a function of them

$$N_{t+1} = N_t \left( \frac{g \, 1.25 R_t / N_t}{f + 1.25 R_t / N_t} \right)$$

$$+ \; 0.4 N_t \left( \frac{a \, 1.25 R_t / N_t}{b + 1.25 R_t / N_t} \right). \tag{8.12}$$

To calculate the equilibrium population (carrying capacity) $N_{eq}$ as a function of rainfall and the parameters, we assume that $R_t = R$, a constant, set $N_t = N_{t+1} = N_{eq}$, and do the necessary algebra. The result is a quadratic equation whose first root is the positive equilibrium population

$$N_{eq} = \frac{-b' + \sqrt{(b')^2 - 4 \, a' c'}}{2 a'}, \tag{8.13}$$

where $a' = bf$, $b' = 1.25R (b + f - gb - 0.4af)$, and $c' = (1.25R)^2 (1 - g - 0.4a)$.

## THE MODELS: WHAT IS THE INTENSITY OF POACHING? (THE 1992 QUESTION)

In 1978 the population was expected to continue to increase except under very-low-rainfall conditions, and there was little chance of a population decline. However, surveys conducted from 1978 to 1989 showed that the size of the wildebeest herd stayed essentially constant, and perhaps declined slightly (Figure 8.1).

A major change in the management of the ecosystem took place in 1978, when Tanzania closed its border to Kenya and the Tanzanian economy went into a severe decline. As a result, fuel and vehicles became almost totally unavailable to park staff and antipoaching patrols by park rangers effectively ceased. Parts of the parks became permanently occupied by poachers, and illegal harvesting almost certainly increased. Campbell and Hofer (1995) estimate that about 120 000 wildebeest were illegally harvested each year from the Serengeti herd. To ask how compatible such estimates of harvest are with the model and data we have used, we must add harvesting to the model.

We do this by allowing the population dynamics to include harvesting after 1977. For example, the dynamics in Equation 8.10 become

$$N_{t + 1} = (s_{\text{adult},t})N_t + (s_{\text{calf},t})B_t \qquad \text{if } t < 1977,$$

$$N_{t + 1} = (s_{\text{adult},t})N_t + (s_{\text{calf},t})B_t - h_t \qquad \text{if } t \geq 1977, \quad (8.14)$$

where $h_t$ is the harvest (illegal take) in year $t$.

## THE CONFRONTATION: THE EFFECTS OF RAINFALL

### Logistic Model

We begin with the logistic model, Equation 8.2, with observation uncertainty and use the census data available in 1978. A pseudocode based on Equation 8.4 that can be used to determine the best values of $r$ and $K$ is:

---

Pseudocode 8.1

1. Input census data up to 1978 (means and standard deviations).
2. Input starting estimates of the parameters $r$, $K$, and $N_1$.
3. Find the values of the parameters that minimize the negative log-likelihood by these steps:
   (a) Predict values of $N_t$ from Equation 8.2.
   (b) Calculate the negative log-likelihood using Equation 8.4 for years in which census data are available.
   (c) Sum the negative log-likelihoods over all years.
   (d) Minimize the total sum of the negative log-likelihoods over $r$ and $K$.

---

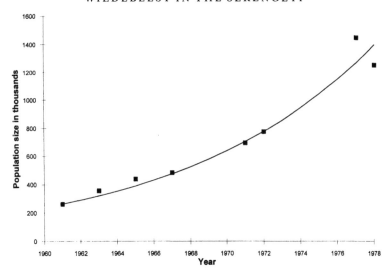

FIGURE 8.4. Population size based on the best-fit logistic model to wildebeest abundance estimates through 1978.

When implementing this pseudocode, population size and all other parameters are constrained to be positive. A computer program based on Pseudocode 8.1 leads to best-fit parameters $r = 0.10$ and $K = 3.5 \times 10^9$, and an excellent agreement between the data and the prediction (Figure 8.4). The logistic model obviously buries many biological details but the fit between the model and the data is excellent.

However, the estimate of carrying capacity is beyond all bounds of reasonable expectation. Why is this the case? Looking carefully at Figure 8.4 provides some of the answer: the best fit to the data is exponential growth—the data provide no real information about $K$ except that it must be large enough that there was no slowdown in increase over the range of population sizes from 1960 to 1978. For the purposes of determining carrying capacity, the census data are uninformative. This conclusion is emphasized if we study the contours of the joint likelihood between $r$ and $K$ (Figure 8.5). Although $r$ is well defined, $K$ is totally unde-

193

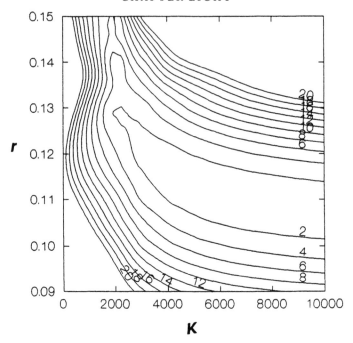

FIGURE 8.5. Contours of $r$ and $K$ for the logistic model.

fined. Thus, this model does not allow us to understand what would happen to the population in the long term, and especially what would happen if rainfall changes. The fit to the data is so good that adding rainfall to the model would be unlikely to be helpful, since there really cannot be any information in the data relevant to rainfall. We conclude, however, that over the range of population sizes seen up to 1978 there is no evidence that the population growth rate was slowing.

### Life History Model

Given the data available in 1978, there is a better chance of understanding what is likely to happen if rainfall decreased after 1978 if we use the information contained in the calf survival and adult survival data. To do this, we confront the life history model (with parameters $a$, $b$, $f$, and $g$) with the census, calf survival, and adult mortality data using a pseudocode such as:

---

Pseudocode 8.2

1. Input rainfall, census, calf survival, and adult mortality data and $N_1$ up to 1978.
2. Input starting estimates of the parameters $a$, $b$, $f$, and $g$.
3. Find the values of the parameters that minimize the negative log-likelihood by
   ($a$) predicting the values of $N_t$ and calf and adult survival, from Equation 8.10;
   ($b$) calculating the negative log-likelihood for the census data using the single-year terms in Equation 8.11;
   ($c$) summing the negative log-likelihoods over all years;
   ($d$) minimizing the total sum of the negative log-likelihoods.

---

As before, the predicted population and all parameters must be constrained to be positive. Because the adult survival data are given as mortality per month of dry season, we assumed that all mortality takes place in four dry season months, which means that the monthly mortality rate $M_{\text{adult},t} = 1 - s_{\text{adult},t}^{1/4}$. We constrained $b \geq 0.1$ to prevent some numerical problems with the equilibrium calculations in Equation 8.13.

The agreement between this model and the census data (Figure 8.6) is just about as good as for the logistic model. However, in this case we also estimate calf and adult survival rates. Thus, we are both introducing additional parameters and seeking additional information. For example, 1968 and 1969 were dry years with less food per animal and higher adult mortality; 1971 and 1972 were wetter years with more food per animal and lower adult mortality. On the other hand, calf survival is relatively constant over the values of food that were encountered (Figure 8.3).

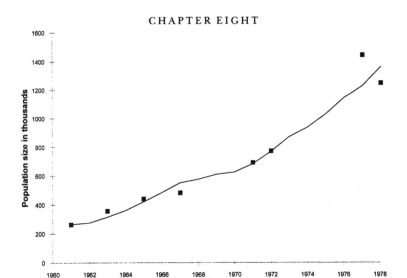

FIGURE 8.6. The predicted trajectory of wildebeest abundance based on the life history model.

How will the future population size be affected by rainfall? Using Equation 8.10 or 8.12 and estimated values of $a$, $b$, $g$, and $f$, the estimated equilibrium population size with 150 mm of dry season rainfall is about 1.8 million, slightly higher than the 1977 and 1978 censuses. Thus the simple prediction is that if rainfall returned to the 150 mm average, the population would stabilize about where it was in the late 1970s. How much confidence do we have in this prediction?

We can explore confidence in the predicted equilibrium population $N_{eq}$, which we identify as the carrying capacity, by computing the likelihood profile on the equilibrium population. Since $N_{eq}$ is not a parameter of the model, we cannot do a direct search over different values of $N_{eq}$ and find the maximum likelihood estimate with all other parameters free. Rather, we must constrain the estimation procedure to find the values of $a$, $b$, $f$, and $g$ that maximize the likelihood, given a specific equilibrium population. This is implemented by defining a target equilibrium population $N_{target}$ and a penalty

$$P(N_{eq}, N_{target}) = \frac{(N_{eq} - N_{target})^{\gamma}}{M}, \qquad (8.15)$$

where $\gamma$ and $M$ are parameters used to set the size of the penalty. We then maximize the sum of the three likelihoods plus the penalty. In effect, we find the best fit constrained so that the equilibrium population is equal to (or very close to) $N_{target}$. A pseudocode to calculate the likelihood profile for $N_{eq}$ is:

---

Pseudocode 8.3

1. Input rainfall, census, calf survival, and adult mortality data.
2. Input starting estimates of the parameters $a$, $b$, $f$, and $g$.
3. Input a range of values of $N_{target}$ over which we search.
4. Loop over the different values of $N_{target}$, and for each value find the values of the parameters that minimize the sum of the negative log-likelihoods of the different data plus the penalty function.
5. Plot or tabulate the negative log-likelihood versus the value of $N_{target}$.

---

In addition to ensuring the constraints described before, the penalty function must have enough weight that $N_{eq}$ is very close to $N_{target}$. By numerical experimentation, we found that $\gamma = 2$ and $M = 100$ worked well, and the nonlinear minimizer converged rapidly.

From the likelihood profile for $N_{eq}$, if dry season rainfall is 150 mm (Figure 8.7), we see that the likelihood is best in the range 1.5 to 2.5 million, but the 95% confidence interval (minimum negative log-likelihood plus 1.92) goes as high as 6 million. Thus, although the life history model combined with calf survival and adult mortality data pro-

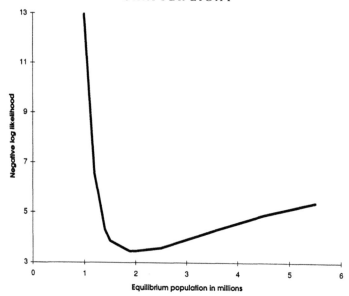

FIGURE 8.7. The negative log-likelihood profile of equilibrium population size for 150 mm of dry season rain.

vides more information about the long-term consequences of rainfall, we cannot exclude considerably larger populations than have been observed. However, the model does exclude a population below 1 million, which was the principal concern in 1978.

In summary, the data available in 1978 indicated that the wildebeest population was unlikely to decline if rainfall returned to 150 mm in the dry season, but the data were insufficient to gauge with any accuracy what level the population might reach.

## THE CONFRONTATION: THE EFFECTS OF POACHING

Now let us turn to the second question. Equation 8.12 with specified values of $a$, $b$, $f$, and $g$ (estimated, using data available in 1978) leads to predictions (Figure 8.8) of the values of population size, and calf and adult survival from

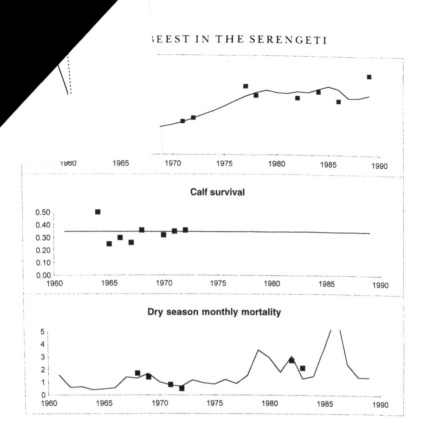

FIGURE 8.8. The predicted and observed data (population size, calf survival, and dry season monthly mortality rates) through 1990, based on parameters estimated through 1978.

1978 onwards. The predictions capture the leveling off of population size, but at a higher level than actually observed. Could this difference be due to poaching?

We use all the data and Equation 8.13 with the annual removals $h_t = h$, a constant. With the addition of $h$, there are five parameters ($a$, $b$, $f$, $g$, and $h$). The minimum negative log-likelihood is $-12.81$, with an estimated kill of 38 000 individuals/year. The minimum negative log-likelihood for a model without harvest (i.e., $h = 0$) is $-11.86$. Since the difference between these two values is less than 1.92 (the critical value for the $\chi^2$ distribution), we conclude

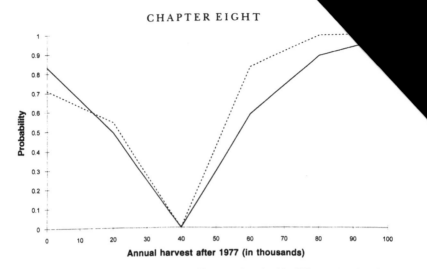

FIGURE 8.9. The chi-square probability associated with different levels of harvest after 1977. The solid line is for the adult mortality relationship Equation 8.9. The dashed line is the result when it is assumed that dry season monthly adult mortality is linearly related to food per animal.

that adding poaching (the parameter $h$) does not lead to a significantly better explanation of the data.

A likelihood profile of harvest gives information on how compatible different levels of harvest are with the model and data (Figure 8.9). The best estimate of harvest is $h = 40\ 000$ (using 20 000 increments), and at $h = 0$ the $\chi^2$ probability is about 0.83, confirming the earlier comparison of negative log-likelihoods that showed that adding poaching to the model did not improve the goodness of fit to the model at the 0.05 level. The right-hand side of the solid line in Figure 8.9 shows that harvests of 80 000 or more are largely incompatible with the model and the data, because if the harvest were that intense, the population would have declined.

Figure 8.9 represents our best estimate of what the harvest could be, given that the model and data are true. However, a major factor determining the population dynamics in the model is adult mortality, which is apparently much more sensitive to food availability than calf survival. All predictions presented so far have been based on the relationship in Equation 8.9 between food per animal and adult mortality, and the estimates of $g$ and $f$ are based on only six data points. An alternative relationship is one in which monthly dry season survival of adults is linearly related to food per animal. If we make this change in the model, we obtain almost identical results (Figure 8.9 dashed line). The likelihood profile of harvest tells the same story. This gives some confidence in our conclusions. Thus, if the data and model are correct, harvests of 120 000 each year after 1978 are incompatible with the census data.

There are at least two ways in which larger harvests could arise. First, the level of poaching may have increased dramatically in the late 1980s as the economy recovered. When antipoaching patrols resumed, many poacher camps were destroyed and hundreds of poachers were arrested, but the sudden availability of fuel and vehicles may have prompted the development of a commercial meat market for poached game, whereas before, limitations on transportation had kept poaching for meat confined to subsistence users. Thus, rather than poaching having been constant since 1978, as assumed, it is possible that it accelerated dramatically in the last few years. Second, the mortality data may be fundamentally too high. The pregnancy, calving, and calf survival rates are quite reliable. In the wet season, at the time of the aerial census, it is easy to count the proportion of the total population that is yearlings. This number has been quite stable, indicating reasonably stable recruitment. The adult survival data, however, are derived from small samples in a few places (and only for a total of six years). If there is a significant error in our assumptions, it is most likely that the mortality data are incorrect.

## IMPLICATIONS

We still cannot estimate the carrying capacity for wildebeest, or how rainfall affects it, with much accuracy. We did show, however, that if rainfall drops to 150 mm/year, a lower limit of the population is about 1 000 000 individuals. We also showed that the estimate of carrying capacity depends to a large extent on the estimates of adult mortality, which are the "weak link" in the chain of logic.

Alfred Lotka (of Lotka-Volterra fame) used data on the human population of the United States from 1790 to 1910 to estimate the parameters of the logistic equation, and concluded that "if the population of the United States continues to follow this growth curve in future years, it will reach a maximum of some 197 million souls by the year 2060." In fact, this level (197 million) was crossed around 1970. Tuckwell and Koziol (1992) fitted population data from 1950 to 1985 to the logistic growth curve. Their model accurately estimated world population in 1992 (about 5.5 billion), and they estimated that the carrying capacity of the world is 23.8 billion and that this will be achieved by 2250. The experience in this chapter suggests that estimating the carrying capacity of the world from data corresponding to the likely "exponential" phase of population growth is a fruitless activity (also see Cohen 1995). Pulliam and Hadad (1994) review human population growth, the carrying capacity concept, and the role for ecologists (and thus ecological detectives) in the human population problem.

In this analysis of Serengeti wildebeest, we used several sources of diverse data in a single, unified framework for modeling, estimating parameters, and comparing hypotheses. By dealing with all data (abundance estimates, calf survival, and adult survival) simultaneously, we could focus on uncertainty in one parameter such as harvest, while admitting uncertainty in the relationship between food per animal and calf survival and adult mortality. We constantly traversed the path between the models and the data, using experience with one to improve investigations of the other.

# The Confrontation: Bayesian Goodness of Fit

## WHY BOTHER WITH BAYESIAN ANALYSIS?

The answer is this: because we often have prior information that is valuable and should not be lost in an analysis. For example, Reader et al. (1994) describe an intercontinental study of plant competition which involved *Poa pratensis* in twelve different communities. Suppose that subsequent to the study, we wanted to model the plant dynamics in one of the communities. Should we discard relevant information from the other eleven? That seems foolish, but a method for incorporating the previous information is needed, and Bayesian methods provide a framework for using prior information. Stow et al. (1995) proposed that some of the debate concerning the appropriate description of consumer-resource interactions (especially the notion of ratio dependence) can be resolved by using Bayesian methods.

Furthermore, we analyze ecological data to determine the relative probability of competing hypotheses. At the end of a scientific paper, we want to be able to say how well the data support each alternative hypothesis, given all the available data. Using all the available data means not only using the results of our experiment, but the results of any previous experiments. Bayesian methods produce estimates of the probabilities of alternative hypotheses based on all the data and this is the goal of science (see Chapter 2).

For example, if there are two competing hypotheses, $H_1$ and $H_2$, and our results show an 80% probability $H_1$ is true and a 20% probability $H_2$ is true, we could stop there. However, someone has to combine the results of all the experiments to determine the probabilities of $H_1$ and $H_2$ not only given our experiment, but including previous work, and it would seem rather pointless to report our results without making reference to other experimental results.

Bayes' theorem provides a simple way to use all possible information. The starting point is Equation 3.9, in which the event $A$ is the data and the event $B$ is hypothesis $H_i$; we replace $\Pr\{A|B\}$ with the likelihood $\mathscr{L}\{\text{data}|H_i\}$ of the data given the hypothesis, and $\Pr\{B\}$ with the prior probability $\text{Prior}\{H_i\}$ assigned to the hypothesis, to obtain

$$\Pr\{H_i|\text{data}\} = \frac{\mathscr{L}\{\text{data}|H_i\}\,\text{Prior}\{H_i\}}{\Pr\{\text{data}\}}.$$ 

(9.1)

Here $\Pr\{H_i|\text{data}\}$ is the probability of the hypothesis, given the data (this is also known as the posterior probability). The prior probability of $H_i$ summarizes what we know before the experiment; it is the posterior probability emerging from the previous experiment.

The numerator is the joint probability of the data and $H_i$. The denominator must be the sum of such joint probabilities, summed over all possible hypotheses. Thus Bayes' theorem is also sometimes written as

$$\Pr\{H_i|\text{data}\} = \frac{\mathscr{L}\{\text{data}|H_i\}\text{Prior}\{H_i\}}{\sum_j \mathscr{L}\{\text{data}|H_j\}\text{Prior}\{H_j\}}.$$ 

(9.2)

For example, assume that in the contest between hypothesis $H_1$ and $H_2$, we conclude that $H_1$ was four times more consistent with the experimental results than $H_2$. This result alone would suggest an 80% probability of $H_1$ and a 20% probability of $H_2$. Suppose that a previous experiment resulted in a 60% probability of $H_1$ and a 40% probability of $H_2$. We treat the previous experiment as the prior and use Bayes' theorem to obtain nents}

$$\frac{\times\ 0.6}{+\ 0.2\ \times\ 0.4} = 0.857, \tag{9.3}$$

he only two hypotheses, then $Pr\{H_2|$
$= 1 - Pr\{H_1|\text{both experiments}\} =$
numerate in 9.3 is the relative likelihood of $H_1$
and the denominator is the probability of the data, which
assures that the total posterior probabilities add up to 1.0.
Because of the prior likelihoods of the two hypotheses, the
second experiment provides better discrimination between
the two hypotheses (this is not always the case).

We formalize this idea a bit more by dividing numerator
and denominator in Equation 9.1 by the likelihood of the
data, given $H_i$:

$$Pr\{H_i|\text{data}\}$$
$$= \frac{Prior\{H_i\}}{\sum_j (\mathcal{L}\{\text{data}|H_j\}/\mathcal{L}\{\text{data}|H_i\})\ Prior\{H_j\}}. \tag{9.4}$$

Thus, the posterior probability of hypothesis $H_i$ is the prior
probability divided by a weighted sum of the prior proba-
bilities. The ratio $\mathcal{L}\{\text{data}|H_j\}/\mathcal{L}\{\text{data}|H_i\}$ is called the "odds
ratio." A good experiment is one in which this ratio is small,
except for one of the competing hypotheses; a bad experiment is one in which this ratio is close to 1 (why does this
make the experiment bad?).

Bayes' theorem, which has been known for two hundred
years, is sometimes called by the intuitive name "inverse
probability" (Jeffreys 1948). The important point is that discriminating between competing hypotheses depends not
only on the experimental results, but also on the prior probabilities of the hypotheses. It is quite dangerous to proceed
without considering previous work—indeed, one of the fundamental elements of the scientific method is to use prior
information. In statistics, a long and sometimes vituperative

205

debate about the appropriateness of Bayesian analysis
tinues. Samaniego and Reneau (1994) provide a good sta
ing point to learn about this debate and potential resolu-
tions to it. On the other hand, in science—particularly
ecology—methods that do not allow us to incorporate prior
information seem to miss the mark considerably.


## SOME EXAMPLES

We begin with two examples with discrete and complementary hypotheses.

### *The Bayesian Squirrel*

Imagine a squirrel that buries all its food in one of two
different and large locations. It traditionally buries food in
location 1 with frequency $p_1$ and in location 2 with frequency $p_2 = 1 - p_1$. If the squirrel spends a day searching
location $i$ and it buried food there, there is a chance $s_i$ that
it will find food on that day. For simplicity, we assume that
the chance of finding food is independent of how many
times the location has been searched (at the end of this
chapter, you will know how to modify this assumption).

Winter has started; where should the squirrel look? To
answer this question, note that $s_i$ is really the conditional
probability of finding food, given that it is there. Thus, the
product $p_i s_i$ is the probability that there is food in location $i$
and the squirrel finds it. So it makes sense that it should
search the location in which $p_i s_i$ is largest. Suppose, for the
sake of argument, that this is true in location 1. The squirrel
goes there and does not find food today. Should it search
location 1 tomorrow or switch to location 2?

The answer to this question requires a Bayesian computation. We set

$p_1' = \Pr\{$food is in location 1 | squirrel searched
there and did not find it$\}$. (9.5)

To use Bayes' theorem, recognize that the numerator should be the probability that the food is in location 1 and is not found, and that the denominator should be the probability that the food is not found, so that

$$p_1' = \frac{p_1(1 - s_1)}{p_1(1 - s_1) + p_2} \,. \tag{9.6}$$

The first term in the denominator is the probability that food is present in location 1 and not found; the second is the probability that food is in location 2 and not found when location 1 is searched (this probability is 1). If there are only two hypotheses, then $p_2' = 1 - p_1'$.

After an unsuccessful search, the probability that food is in either location is "updated" using Equation 9.6 or the equivalent if location 2 is searched,

$$p_2' = \frac{p_2(1 - s_2)}{p_2(1 - s_2) + p_1} \,. \tag{9.7}$$

These are the prior probabilities of location for the next day. That is, after the updating, we replace $p_1$ by $p_1'$ and $p_2$ by $p_2'$. By doing this, we continually incorporate all the information acquired previously (the locations of unsuccessful search). Starting with $p_1 = 0.7$, $p_2 = 0.3$, $s_1 = 0.8$, and $s_2 = 0.4$ (can you explain why $p_1 + p_2$ must sum to 1 but $s_1 + s_2$ need not?), we obtain the results shown in Table 9.1.

The results in this table are completely deterministic: for the same starting conditions, we always obtain the same sequence of updated probabilities and the same recommendations about where to search. Here the "decision" of the squirrel, which location to search, is based on the highest posterior chance of finding food. Notice two features. First, the best site to search flips back and forth between the two. Second, this calculation provides information on the location of the food, conditioned on unsuccessful search. An additional computation (which you should do) is needed to find the probability of successful search. In this case, the

TABLE 9.1. Probability of location of food sought by the Bayesian squirrel.

| Day | $p_1$ | $p_2$ | $p_1 s_1$ | $p_2 s_2$ | Location for search |
|-----|-------|-------|-----------|-----------|---------------------|
| 1 | 0.7 | 0.3 | 0.56 | 0.12 | 1 |
| 2 | 0.318 | 0.682 | 0.255 | 0.273 | 2 |
| 3 | 0.438 | 0.562 | 0.35 | 0.225 | 1 |
| 4 | 0.135 | 0.865 | 0.108 | 0.346 | 2 |
| 5 | 0.206 | 0.794 | 0.165 | 0.318 | 2 |
| 6 | 0.302 | 0.698 | 0.241 | 0.279 | 2 |
| 7 | 0.419 | 0.581 | 0.335 | 0.233 | 1 |
| 8 | 0.126 | 0.874 | 0.101 | 0.350 | 2 |
| 9 | 0.194 | 0.806 | 0.155 | 0.323 | 2 |

chance that the animal actually has to go until day 10 is about $10^{-5}$.

*Rumpole the Bayesian*

In 1990 an Englishman was sentenced to sixteen years in jail for raping three women. His conviction was based in large part on DNA fingerprinting of samples taken from the scene of the crime. Expert witnesses reported there was a 1 in 3 000 000 chance that a match between the convicted man and the samples would have occurred by chance alone. The simple implication of this result is that there is only a 1 in 3 000 000 chance that the man was not guilty—far beyond a shadow of a doubt.

If we apply Bayes' theorem to this problem, the competing hypotheses are that the man is guilty or that he is innocent. We want to calculate the posterior probability Pr{innocent} that he is innocent. We need to know the likelihood $\mathscr{L}${DNA match | innocent} of the match between his DNA and those of samples from the scene of the crimes. This is 1 in 3 000 000. Most importantly, we need the prior probability Prior{innocent} that he is innocent. How do we find the prior probability? This depends very much on how this individual man was chosen for DNA testing. Imagine, for instance, that there existed a national DNA data base on all men in England, and the accused was found by searching the data base for the best match (this was not the case but is used as an illustration). In that case, the prior probability that he was guilty would be 1 out of the number of men in the data base, which could be perhaps 10 million. The prior probability he was innocent is thus 9 999 999/10 000 000. If we then apply Bayes' theorem to calculate the probability that the man is innocent, we obtain

Pr{innocent | DNA match}

$$= \frac{(1/3\ 000\ 000)\ (9\ 999\ 999/10\ 000\ 000)}{(1/3\ 000\ 000)\ (9\ 999\ 999/10\ 000\ 000)\ +\ 1/10\ 000\ 000}$$

$$= 0.77. \tag{9.8}$$

The numerator is the joint probability that he is innocent and a match is obtained, and the denominator is the probability that a match is obtained. It is the sum of the probability that he is innocent and a match is obtained and the probability that he is guilty and a match is obtained. Note that we assume that the probability that a match is obtained given that he is guilty is 1. Since there are only two hypotheses, the probability that he is guilty given the data is 0.23.

The intuition underlying this result is that if we search 10 million men for DNA, it is likely that we will find several that match with a probability level of 1 in 3 million. However, the key is that the posterior probability of innocence depends on the prior probability of innocence as well as the experimental evidence. In this case, the prior probability of innocence changes the chance of innocence from 1 in 3 million to 77 in 100.

On the other hand, if there were only 10 000 men of the right age living in the local area and only these men had their DNA tested, then the posterior probability of innocence drops from 0.77 to 0.003. While this would certainly satisfy the scientific journal criterion of 0.05 or 0.01, the evidence to the jury that there was a 3 in 1000 chance he was innocent, based on the DNA evidence, is quite different from a 1 in 3 million chance. The Bayesian method allows one to incorporate the prior information about the number of men tested.

Once again, we see that discriminating between competing hypotheses depends not only on experimental results, but on the prior probabilities of the hypotheses. Good (1995) illustrates the same point for the forensic question about the likelihood that a man who batters his wife will one day murder her.

*Fisher's Lament*

D. Basu (in Ghosh 1988, Chapter IV) tells a story about a meeting that he, Sir Ronald Fisher, and Professor R. R. Bahadur had in the late 1950s. At that time, Basu was having trouble understanding likelihood and frequentist methods when one has certain knowledge that the parameter of interest lies in a known interval. We take the liberty of converting Basu's recollection into a script. It would probably be best to read this section with a partner; it is irrelevant who plays Fisher and who plays Basu.

We believe that Basu was not trying to make fun of Fisher's theories. The message is that the effort is probably better spent understanding the prior information than convincing yourself that a particular statistical approach is the best one.

> SIR RONALD: Basu, why are you having so much trouble understanding the fiducial logic? [Note: "Fiducial" means based on firm faith or used as a standard of reference in a calculation.]
> BASU: Sir Ronald, allow me to ask. With a sample $x$ of a normal random variable with unknown mean $\mu$ and known variance $\sigma^2$, the fiducial distribution on $\mu$ is a normal distribution with mean $x$ and variance $\sigma^2$, that is, $\mathcal{N}(x,\sigma^2)$.

CONT.

How should we modify the fiducial distribution when we have the sure prior knowledge that $\mu$ lies in the closed interval $[0,1]$?

SIR RONALD (with confidence): All the probability mass of the fiducial distribution $\mathcal{N}(x,\sigma^2)$ that lies to the right of 1 should be stacked on 1. Similarly stack all the probability mass to the left of 0 on 0.

BASU (turning aside): I knew what the reply would be and so am prepared with my next question.

BASU (to SIR RONALD): Sir, consider the situation where the known variance $\sigma^2$ is a very large number. Even before the sample $x$ is observed, it is likely that the value is going to fall outside the interval $[0,1]$ and that we are going to put well over 50% of the fiducial probability mass at the two end points 0 and 1. Thus the mere knowledge that the mean lies in $[0,1]$ makes us mentally prepared to accept the proposition that it is 0 or it is 1.

SIR RONALD (angered at such impertinence): Basu, either you believe in what I say or you don't, but never ever try to make fun of my theories.

## The Masses of Neutron Stars

What, you might ask, is something about neutron stars doing in *The Ecological Detective*? Well, the example is terrific (Finn 1994; Maddox 1994). Furthermore, ecology has commonalities with the earth sciences (Roughgarden et al. 1994) and with astrophysics, as Cowell (1984) eloquently describes:

The last sentence of this quotation [due to Robert Mac-Arthur] brings up one final point I wish to make. "The only real rules in science are honesty and validity of logic." *Experimental* falsifiability is not a rule, it is a tool, like mathematical modeling, statistical inference, a pair of binoculars, or an electron microscope. Observation, logical inference, and plausibility arguments are sometimes as

211

capable of scientific revelation as experiments and statistics. Realistic experiments with primate or bird communities are not very much more feasible than experiments in astrophysics, but our curiosity about stars and starlings is not thereby lessened.

Finn (1994) notes that there have been four observations of binary pulsars (two stars rapidly circling one another; in several cases one of them is thought to be a neutron star) that provide information about the total mass $M$ of the pulsar and the mass of the presumed neutron star. Finn uses these limited data in a Bayesian way to determine the joint probability distribution of the lower and upper limits of the mass distribution of neutron stars. Finn (1994) assumed a uniform prior distribution to determine a Bayesian confidence interval (we describe such a computation below), and also explored the use of different priors in the conclusions about the posterior distributions. Maddox (1994) notes that it "will be interesting to see how quickly the [upper and lower mass] limits close up upon each other as further data accumulate." This is an advantage of the Bayesian method: the posterior that Finn computed is the prior when the next set of data is collected. Maddox also notes, "Meanwhile, it seems inevitable that this example will quickly find its way into some textbooks as an illustration of how inferences can be drawn from a meager collection of data." That prediction is correct.

### Bayesian and Likelihood Methods Are Essentially the Same for Discrete Hypotheses

The first two examples show that as long as we are dealing with discrete hypotheses, such as that food is in location 1 or location 2 or the man is innocent or guilty, there is little difference between treating a problem from a Bayesian perspective or from a likelihood perspective. The one difference arises when there are no previous experimental results on which to base the prior probabilities. We delve into this matter more in this chapter and in the next chapter.

When one lacks previous experimental data, a common practice is to assign all hypotheses equal prior probability. A problem with this approach occurs, however, when two of the hypotheses are similar. For example, as fish stocks are depleted due to exploitation, the consequences of alternative harvesting options often depend a great deal on whether the recruitment to the population of young fish will decline dramatically, slightly, or not at all as the total spawning stock size is reduced. It is common practice to consider that three alternative recruitment hypotheses are

$H_1$: Recruitment will stay roughly constant.
$H_2$: Recruitment will decline linearly as spawning population falls.
$H_3$: Recruitment will show depensation and drop more rapidly than spawning stock declines.

The data available to scientists often are not especially informative on this issue and are equally compatible with all three hypotheses, so that the likelihood of the data, given the hypothesis, is approximately the same for each hypothesis. Therefore, the posterior probability we assign to these three hypotheses will depend primarily on the prior probabilities. However, $H_2$ and $H_3$ often give similar predictions about the expected results of alternative management actions. By assigning the hypotheses equal prior probability, we effectively give $2/3$ probability to the response associated with declining stock size. In fact, we might recognize that the second and third hypotheses are really variations of the theme that recruitment falls with spawning stock. That is, there are really two competing hypotheses:

$H_a$: Recruitment does not decline as stock size does.
$H_b$: Recruitment declines as stock size does.

If this is the case, the best prior assignment of probabilities should be 50% for $H_1$ and 25% for $H_2$ and $H_3$.

There is no easy answer to the question of prior probability assignments. For the recruitment example, one should examine the history of many similar fish stocks and determine how many displayed relatively constant recruitment, recruitment that declined proportional to spawning stock, or depensatory recruitment collapses. The relative frequency of the observed occurrences in other stocks can then serve as the prior probabilities for the problem at hand.

The fact of life is that in order to compute the probability of a hypothesis, given all the data, one must assign each hypothesis a prior probability. Determining the relative probability of alternative hypotheses without assigning prior probabilities means that you are making implicit assumptions. So, we all are Bayesians, whether we like it or not!

There are many excellent texts on Bayesian analysis; these provide an entry into the more complicated methods for assigning prior probabilities. Our favorites are by Jeffreys (1948), DeGroot (1970), Berger (1980), Martz and Waller (1982), and Gelman et al. (1995); we encourage you to look at them.

MORE TECHNICAL EXAMPLES

We now illustrate Bayesian methods through a sequence of examples that show how to construct priors and posteriors and how to use the results.

*Counting Emerging Animals*

Suppose that we are measuring the emergence of animals. This could range from insects emerging from pupal cases to mammals ending hibernation. Assume that the number of emergences per unit time (e.g., insects/day or bears/week) can be modeled by a Poisson distribution with rate parameter $r$. The hypothesis is the value of $r$, which is a continuous variable greater than 0. Thus

Pr{data | hypothesis}

= Pr{$k$ emergences in a single period | emergence rate is $r$}

$$= \Pr\{k|r\} = \frac{e^{-r}r^k}{k!} . \tag{9.9}$$

We also view Equation 9.9 as the likelihood

$$\mathcal{L}\{k|r\} = \frac{e^{-r}r^k}{k!} , \tag{9.10}$$

and use this interchangeably with $\Pr\{k|r\}$. As with discrete hypotheses, we want to use the data to make statements about the likelihood of different values of the emergence parameter. We begin with Bayes' theorem,

Pr{rate of emergence is $r$ | $k$ emergences in a single period}

$$= \frac{\Pr\{k \text{ emergences} | \text{emergence rate is } r\} \text{ Prior } \{r\}}{\Pr\{k \text{ emergences}\}} . \tag{9.11}$$

Equation 9.11 is fundamentally no different from Equation 9.1 or 9.2, but there is an important operational difference between Equation 9.11 and the previous examples. Because the emergence rate $r$ is a continuous variable, we must characterize it by a prior probability density function $f_{\text{prior}}(r)$,

Pr{emergence rate is in the interval $r$ to $r + \Delta r$}

$$= f_{\text{prior}}(r)\Delta r + o(\Delta r). \tag{9.12}$$

As before, and throughout the rest of the chapter, we write

$$\Pr\{\text{emergence rate is approximately } r\} = f_{\text{prior}}(r), \quad (9.13)$$

and $f_{\text{prior}}(r)$ summarizes what we know about the rate of emergence (e.g., from work in other years or from studies in other places). The use of a density function, rather than discrete hypotheses, also means that one has to use calculus when applying Bayes' theorem to compute the posterior

215

probability density $f_{post}(r|k)$ for the emergence parameter, given the data.

The numerator in Equation 9.11 is now

Pr{observe $k$ emergences and emergence rate is approximately $r$}

$$= f_{prior}(r) \frac{e^{-r}r^k}{k!},$$

(9.14)

and the denominator in Equation 9.11 is

Pr{observe $k$ emergences}

$$= \int_0^\infty f_{prior}(r') \frac{e^{-r'}(r')^k}{k!} \, dr'.$$

(9.15)

Note that we choose a different symbol for the integration over all possible values of the emergence parameter. Combining these, the posterior density is

$$f_{post}(r) = \frac{f_{prior}(r) \, e^{-r}r^k/k!}{\int_0^\infty f_{prior}(r') \, (e^{-r'}(r')^k/k!) \, dr'}.$$

(9.16)

Once we choose a prior density, we can evaluate the posterior density. What prior should we use? First, we could assume that we know nothing about the value of $r$, except that it must be non-negative, and we could choose the uniform prior for which $f_{prior}(r) = 1$. If $r$ is unbounded, this prior density cannot be normalized, so that the prior is really not a probability density function. Such prior densities are called *improper*. If we choose the uniform prior for $r$, Equation 9.16 becomes

$$f_{post}(r) = \frac{e^{-r}r^k/k!}{\int_0^\infty (e^{-r'}(r')^k/k!) \, dr'},$$

(9.17)

which we recognize as a gamma density with parameters $k+1$ and 1. This suggests something: perhaps we should model the prior as a gamma density. That is, suppose we assume

$$f_{\text{prior}}(r) = \frac{a^n}{\Gamma(n)} e^{-ar} r^{n-1}. \tag{9.18}$$

This choice of prior means that we summarize the prior information in that the prior mean of the emergence rate is $n/a$ and the prior coefficient of variation of the emergence rate is $1/\sqrt{n}$. Using the prior Equation 9.18 in Equation 9.11 gives

$$f_{\text{post}}(r)$$
$$= \frac{[a^n/\Gamma(n)] \, e^{-ar} \, r^{n-1} \, [e^{-r} r^k/k!]}{\displaystyle\int_0^\infty [a^n/\Gamma(n)] \, e^{-ar'} \, (r')^{n-1} \, [e^{-r'}(r')^k/k!] \, dr'}. \tag{9.19}$$

Although this expression is complicated, it can be simplified. First, note that the constant terms $a^n$, $k!$, and $\Gamma(n)$ cancel. Second, note that the exponential and polynomial terms combine to give

$$f_{\text{post}}(r) = \frac{e^{-(a+1)r} \, r^{n+k-1}}{\displaystyle\int_0^\infty e^{-(a+1)r'} \, (r')^{n+k-1} dr'}. \tag{9.20}$$

That is, the posterior density is another gamma density with parameters $n + k$ and $r + 1$. The gamma density is called the *conjugate prior* for the Poisson process because if the prior is a gamma, then the posterior is also a gamma but with changed parameters. We do not need to compute the probability density as data are collected, we only need to change the parameters of the density.

For an arbitrary prior density, $f_{\text{prior}}$, we usually must resort to numerical evaluation of Bayes' theorem. To do this, we

217

specify minimum $r_{min}$ and maximum $r_{max}$ values of $r$ and replace the integral by a sum that runs from $r_{min}$ to $r_{max}$ in steps $\Delta r$ (which we also specify). We will use the notation $\sum_{r' = r_{min}(\Delta r)}^{r' = r_{max}}$ to indicate that a sum runs from $r_{min}$ to $r_{max}$ in steps of $\Delta r$. Thus, the denominator in Equation 9.16 becomes

$$\sum_{r' = r_{min}(\Delta r)}^{r' = r_{max}} f_{prior}(r') \frac{e^{-r'} (r')^k}{k!} \Delta r. \tag{9.21}$$

Note, however, that the denominator in Equation 9.16, or its discrete version Equation 9.21, really serves to ensure that the posterior density is properly normalized. A pseudocode for computing the posterior density is:

---

Pseudocode 9.1
1. Specify the data ($k$), the prior density $f_{prior}(r)$, the minimum $r_{min}$ and maximum $r_{max}$ values of $r$, and the step $\Delta r$.
2. Use Equation 9.21 to compute the denominator in Equation 9.16.
3. Compute the posterior density by cycling over the values of $r$, from $r_{min}$ to $r_{max}$ in steps of $\Delta r$, and using Equation 9.16.

---

Employing this algorithm, we can generate information about the posterior density of the rate parameter (Figure 9.1). Suppose that the data are $k = 4$ counts in one interval of time. If we adopt the prior given in Equation 9.18 by changing the values of $n$ and $a$, we can change the amount of initial information contained in the prior.

We encourage you to try the following exercise. Suppose that a uniform prior is assumed and that in the first observation period, four emergences are observed. Compute and plot the posterior that then becomes the prior for the next observation period. Assume that in the next observation pe-

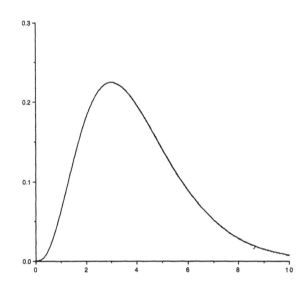

**Emergence rate**

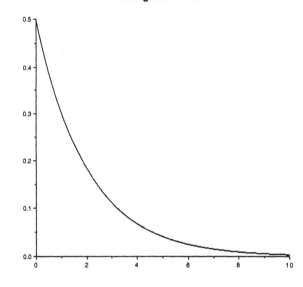

**Emergence rate**

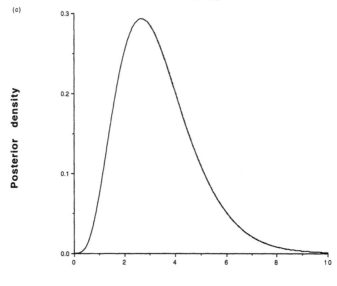

(c)

Posterior density

Emergence rate

FIGURE 9.1. Bayesian posterior densities on the rate parameter of the Poisson process when the data are $k = 4$ counts in 1 time period. (a) When the prior density is the improper uniform $f_{\mathrm{prior}}(r) = 1$, the posterior is a gamma with parameters $n = 4$ and $a = 1$. (b) A prior density that is a gamma function, with parameters $n = 1$ and $a = 0.5$, corresponds to prior information about the rate parameter suggesting that the mean is 2 and the coefficient of variation is 100%. (c) The posterior density, updated from the prior shown in (b), now has parameters $n = 1 + k = 5$ and $a = 0.5 + 1 = 1.5$.

riod three emergences are observed. Compute and plot the new posterior.

*Return to Random Search.* In Chapter 3, we showed (Box 3.2) that if the probability that a predator finds food in search time $t$ is $1 - e^{-ct}$, where the search parameter $c$ is fixed, then search is memoryless, but that if $c$ is uncertain and thus has a distribution, this need not be the case (Equation 3.86 and following). We assumed that $c$ had a gamma density with parameters $n$ and $a$, which we can now recognize as a prior density. We are interested in the posterior density for values of $c$, given unsuccessful search $f_{post}(c|$ unsuccessful search), which is

$$f_{post}(c \mid \text{unsuccessful search})$$

$$= \frac{e^{-ct}\,[a^n/\Gamma(n)]\,e^{-ac}c^{n-1}}{\displaystyle\int_0^\infty e^{-c't}\,[a^n/\Gamma(n)]\,e^{-ac'}c'^{n-1}\,dc'}\,,$$

$$(9.22)$$

and which is very similar to Equations 9.19 and 9.20. Thus, if the prior density for $c$ is a gamma with parameters $n$ and $a$, the posterior density, given unsuccessful search in $t$ units of time, is also a gamma with $n$ and $a + t$. Although the coefficient of variation of $c$ does not change, the mean decreases from $n/a$ to $n/(a + t)$ after unsuccessful search. This represents learning: the predator's view of the world, summarized in the likelihood of different values of $c$, changes when the search is unsuccessful.

*Sampling the Pistachio Tree.* Suppose that we are sampling a tree to determine the level of infestation of nuts by insect pests. The random variable is the fraction of nuts that are infested, which we denote by $P$, with particular value $p$. In this case, too, the hypotheses—different values of $p$ ranging from 0 to 1—are continuous, so we denote the prior density by $f_{prior}(p)$.

If we sample $S$ nuts and $i$ of them are infested, the likelihood is

$$\mathscr{L}\{i|S,p\} = \binom{S}{i}\, p^i(1 - p)^{S-i}.$$

$$(9.23)$$

Applying Bayes' theorem, the posterior density is

$$f_{post}(p) = \frac{f_{prior}(p)\,\dbinom{S}{i}\,p^i(1 - p)^{S-i}}{\displaystyle\int_0^1 f_{prior}(p')\,\dbinom{S}{i}\,(p')^i(1 - p')^{S-i}\,dp'}\,.$$

$$(9.24)$$

Because $p$ is a fraction, it can only range between 0 and 1; hence those are the limits of integration in Equation 9.24. We now need to choose the prior. One choice is the uniform density, which is a proper prior density in this case because $p$ ranges between 0 and 1. Alternatively, in analogy to the Poisson-gamma case, we might pick a prior density that has the same mathematical form as the likelihood. That is, we assume

$$f_{prior}(p) = c_{norm} p^A (1 - p)^B. \tag{9.25}$$

In this equation, $c_{norm}$ is a normalization constant that is required to ensure that $\int_0^1 f_{prior}(p)\, dp = 1$,

$$c_{norm} = \frac{1}{\displaystyle\int_0^1 (p')^A (1 - p')^B \, dp'}. \tag{9.26}$$

Using this choice of prior in Bayes' theorem leads to the posterior density

$$
\begin{aligned}
f_{post}(p) &= \frac{c_{norm}\, p^A (1 - p)^B \binom{S}{i} p^i (1 - p)^{S-i}}{\displaystyle\int_0^1 c_{norm}\, (p')^A (1 - p')^B \binom{S}{i} (p')^i (1 - p')^{S-i}\, dp'} \\[2ex]
&= \frac{p^A (1 - p)^B\, p^i (1 - p)^{S-i}}{\displaystyle\int_0^1 (p')^A (1 - p')^B\, (p')^i (1 - p')^{S-i}\, dp'}.
\end{aligned}
\tag{9.27}
$$

The second equality in Equation 9.27 follows because $c_{norm}$ and $\binom{S}{i}$ are constants that can be canceled in both numerator and denominator. The denominator in Equation 9.27 is a constant, determined so that $f_{post}(p)$ is normalized to 1 (Box 9.1).

---

BOX 9.1

THE BETA DENSITY

The integral in Equation 9.26 is related to the *beta function* (Abramowitz and Stegun 1965, 258)

$$B(z,w) = \int_0^1 t^{z-1}(1 - t)^{w-1}dt,$$

with the properties that $B(z,w) = B(w,z)$ and

$$B(z,w) = \frac{\Gamma(z)\Gamma(w)}{\Gamma(z + w)} .$$

Thus, we can write the normalization constant in terms of the gamma function,

$$c_{norm} = \frac{\Gamma(A + B + 2)}{\Gamma(A + 1)\Gamma(B + 1)} .$$

For this reason, the prior density Equation 9.25 is called a beta density and this model is sometimes called the "beta-binomial model." Crowder (1978) gives a nice application in a study of the germination of seeds and shows how this Bayesian approach can be combined with other statistical tools.

---

From Equation 9.27, we see that if the prior is proportional to $p^A(1 - p)^B$, then the posterior will be proportional to $p^{A+i}(1 - p)^{B+S-i}$; the posterior has the same form as the prior, with "updated" parameters and normalization constant.

Once we have the posterior density, we can compute the *Bayesian confidence interval* in the following manner. First, let $p^*$ be the value of $p$ that maximizes $f_{post}(p)$. We can determine the symmetric (about $p^*$) $\alpha$ confidence level (e.g., $\alpha = 0.9, 0.95, 0.99$) by finding the value $p_\alpha$ such that

$$\int_{p^* - p_\alpha}^{p^* + p_\alpha} f_{\text{post}}(p) \, dp = \alpha.$$

(9.28)

The confidence interval is then $[p^* - p_\alpha, \ p^* + p_\alpha]$. To actually implement this computation, we need to replace the integral by a sum. Using the notation $\Sigma_{p = p_{\min}}^{p_{\max}}(\Delta p)$ to denote a sum that goes from $p = p_{\min}$ to $p_{\max}$ in steps of $\Delta p$, the approximation of Equation 9.28 is

$$\sum_{p = p^* - p_\alpha(\Delta p)}^{p^* + p_\alpha} f_{\text{post}}(p) \Delta p.$$

(9.29)

Other, asymmetric confidence intervals could also be picked (we use one in the next example).

Similarly, the denominator in Equation 9.27 becomes

$$\sum_{p' = 0(\Delta p')}^{1} (p')^A (1 - p')^B (p')^i (1 - p')^{S-i} \Delta p' .$$

(9.30)

The pseudocode for implementing these ideas is very similar to the one used for the Poisson-gamma case, so we leave it to you to write.

The prior probability density allows us to summarize previous information. For example, suppose that we thought there was a 50-50 chance that a given nut is infested. We might pick $A = B = 1$; this gives a very wide prior density (the curve marked "prior" in Figure 9.2a). Even a little bit

---

FIGURE 9.2. The Bayesian posterior for the parameter in a binomial distribution. The prior probability density is $f_{\text{prior}}(p) = c_{\text{norm}} p^A (1 - p)^B$. (a) When $A = B = 1$, the prior mean is 0.5 and this is also the most likely value. If the data were that of $S = 5$ nuts sampled and $i = 2$ of them were infested, the posterior density has a mean 0.444, a most likely value 0.43, and a 95% confidence interval [0.13,0.73]. (b) When $A = 1$, $B = 0.1$, the prior mean is 0.643, but the most likely prior value of $f$ is 0.91. The same data ($S = 5$, $i = 2$) shift the prior more dramatically than in the previous case. We find that the posterior mean is 0.493, the most likely posterior valley is 0.49, and the 95% confidence interval is [0.18,0.8].

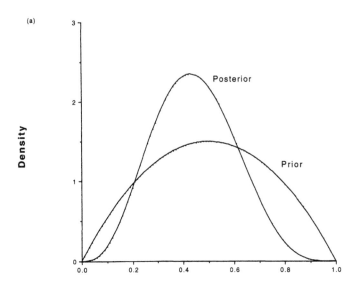

(a)

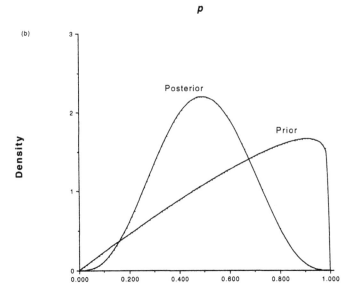

(b)

of data, such as two infested nuts in five nuts sampled has the effect of shifting the peak of the prior and making it less diffuse (the curve marked "posterior" in Figure 9.2a). On the other hand, we might be much more pessimistic about the situation and presume that even though the mean of $p$ is about 0.5, it is highly likely that $p$ is much larger. We can incorporate such information by choosing $A = 1, B = 0.1$ (the curve marked "prior" in Figure 9.2b). The same information ($i = 2, S = 5$) once again changes the prior, but leads to different posterior means, most likely values, and confidence intervals.

We can use this method to explore how our certainty (or lack of it) about the prior value of the parameter affects the conclusions we draw with a little bit more data ($S = 20, i = 3$). For example, we might expect that on average the chance that a nut is infested is 0.5, but could be uncertain about the level of confidence in the prior information. A situation like this can be handled by setting $A = B$ but letting their values vary. As $A$ and $B$ decrease, we are less and less certain about the prior information concerning the chance of infestation (Figures 9.3a–9.3c). In each case, the sampling information ($S = 20$ nuts were sampled and $i = 3$ of them were infested) leads to a posterior that is less diffuse

FIGURE 9.3. Prior and posterior probability densities for the parameter in a binomial distribution for the case in which we expect that the average value of $p$ is 0.5, but have differing levels of confidence about this mean. For all values of $A$ and $B$ used here, the prior mean is 0.5 and the prior most likely value is 0.5. The data are $S = 20$ nuts sampled, and $i = 3$ of the sampled nuts are infested. (a) $A = B = 1$. The posterior mean is 0.208, the most likely posterior value is 0.18, and the 95% confidence interval is [0.01,0.35]. (b) $A = B = 0.5$. The posterior mean is 0.196, the most likely posterior value is 0.17, and the 95% confidence interval is [0,0.34]. (c) $A = B = 0.1$. The posterior mean is 0.184, the most likely posterior value is 0.15, and the 95% confidence interval is [0,0.33]. Note that in no case is the posterior mean as small as the MLE value of $p = 3/20 = 0.15$.

(a)

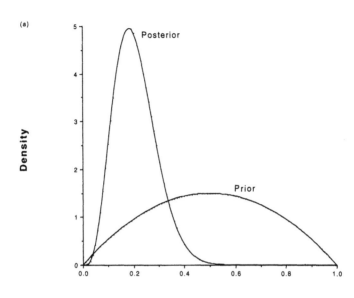

*p*

(b)

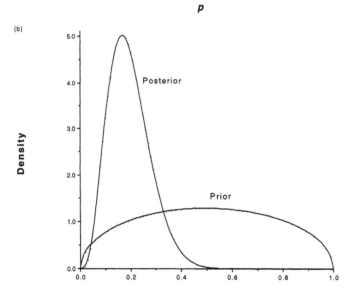

*p*

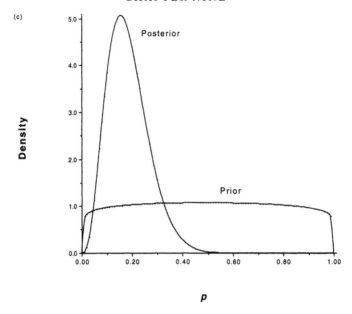

FIGURE 9.3. (*Cont.*)

than the prior. In fact, you may have noted that the posterior densities in Figures 9.2 and 9.3 look very similar. There is a reason for this: as more and more data are collected, the posterior is determined more by the data and less by the prior. The great advantage of Bayesian analysis is that it allows us to incorporate prior information and uncertainty when we have limited data.

*How Many Animals Were Present in a Sampled Region*

Suppose that a region of fixed size, which is closed to immigration and emigration, is sampled over a fixed period, and that animals are removed from the population (by either physical removal or tagging) after capture. If $p$ is the probability of capturing a single animal (assumed to be known and the same for all individuals, but see below) and $N$ animals are initially present, then the number of captures $C$ follows a binomial distribution

$\Pr\{\text{number of captures } C = c \mid N \text{ animals present}\}$

$$= \binom{N}{c} p^c (1 - p)^{N-c}. \tag{9.31}$$

When $N$ is unknown (and thus values of $N > c$ are the hypotheses), we can use Bayesian analysis to make statements about the likelihood of differing values of $N$. We begin by recognizing that Equation 9.31 also defines the likelihood of the data $C = c$, given $N$. We write this as

$$\mathscr{L}\{c|N\} = \begin{cases} \binom{N}{c} p^c (1 - p)^{N-c} & \text{for } N = c, c + 1, c + 2, \ldots, \\ 0 & \text{for other values of } N. \end{cases} \tag{9.32}$$

That is, $N$ is certainly greater than or equal to $c$, and can only take integer values since it is the number of animals present.

Applying the Bayesian procedure, the posterior is

$$f_{\text{post}}(N|c) = \frac{\mathscr{L}\{c|N\} f_{\text{prior}}(N)}{\displaystyle\sum_{N'=c}^{\infty} \mathscr{L}\{c|N'\} f_{\text{prior}}(N')}, \tag{9.33}$$

where we use the notation $f_{\text{post}}(N|c)$ to remind us that the data are $C = c$. We still must choose the prior distribution for $N$. The uniform prior,

$$f_{\text{prior}}(N) = 1, N = c, c + 1, c + 2, \ldots, \tag{9.34}$$

is improper since $\sum_{N=c}^{\infty} f_{\text{prior}}(N) = \sum_{N=c}^{\infty} 1$ is infinite. However, as we shall see, the posterior defined by Equation 9.33 can be normalized even if we use the uniform prior. An alternative uniform prior would limit $N$ to the range 0 to $N_{\text{max}}$.

With the improper prior density in Equation 9.34, the posterior density defined by Equation 9.33 is

$$f_{\text{post}}(N|c) = \frac{\mathscr{L}\{c|N\}}{\displaystyle\sum_{N'=c}^{\infty} \mathscr{L}\{c|N'\}}. \tag{9.35}$$

We can simplify the denominator in Equation 9.35 by using the algebraic identity

$$\sum_{N'=c}^{\infty} \binom{N'}{c} (1 - p)^{N'-c} = p^{-c-1}$$

(9.36)

(Mangel and Beder 1985, 153, Equations 2.19–2.21). The denominator in Equation 9.35 is thus

$$\sum_{N'=c}^{\infty} \mathscr{L}\{c|N'\})$$

$$= \sum_{N'=c}^{\infty} \binom{N'}{c} p^c (1 - p)^{N'-c} = p^{-1},$$

(9.37)

which leads to the posterior density

$$f_{\text{post}}(N|c) = \binom{N}{c} p^{c+1} (1 - p)^{N-c}$$

$$\text{for } N = c, c + 1, \ldots.$$

(9.38)

Now let us consider Bayesian confidence intervals, which we center around the most likely value of $N$, given the data. To find this value, which we denote by $N_{\text{MLE}}(c)$, note that the ratio of two neighboring values (i.e., values of $N$ that differ by 1) of the likelihood is

$$\frac{f_{\text{post}}(N + 1|c)}{f_{\text{post}}(N|c)} = \frac{\binom{N+1}{c}}{\binom{N}{c}} (1 - p)$$

$$= \frac{N + 1}{N + 1 - c} (1 - p).$$

(9.39)

The ratio in Equation 9.39 equals 1 if $N = (c/p) - 1$. Since the MLE value must be an integer, we set

$$N_{\text{MLE}}(c) = \text{Int}\left[\frac{c}{p}\right].$$

(9.40)

We find the symmetric Bayesian confidence interval of probability $\alpha$ around this MLE in a manner similar to the one used for the pistachio example. We seek the smallest $N_\alpha$ so that

$$f_{post}(N_{MLE}(c)) + \sum_{N = N_{MLE}(c) - N_\alpha}^{N_{MLE}(c) + N_\alpha} f_{post}(N|c) \geq \alpha. \tag{9.41}$$

Since we are working with integers, we may not be able to exactly hit the confidence level $\alpha$, and have replaced "$= \alpha$" by "$\geq \alpha$."

An interesting situation, which illustrates the power of the Bayesian method, arises if $c = 0$, i.e., if trapping produces no encounters. Then the MLE value is $N_{MLE} = 0$, and the posterior density Equation 9.38 becomes

$$f_{post}(N \mid \text{no encounters}) = p(1 - p)^N,$$

$$N = 0,1,2, \ldots . \tag{9.42}$$

The Bayesian confidence interval cannot possibly be symmetrical and must be of the form $[0,N_\alpha]$, so that $N_\alpha$ is now the estimate for the maximum number of animals in the region at confidence level $\alpha$. Setting the lower limit of 9.41 to 0, the upper limit to $N_\alpha$, substituting Equation 9.42 into Equation 9.41, and solving gives

$$N_\alpha = \frac{\log(1 - \alpha)}{\log(1 - p)} - 1. \tag{9.43}$$

This equation allows us to predict (Figure 9.4) the number of animals in the region, given unsuccessful trapping but information about $p$.

What would happen if $p$ were also unknown? This means that both the chance of encountering an animal and the number of animals are unknown. In this case, standard MLE methods fail completely. To see this, note that if Equation 9.32 is viewed as a likelihood for both $N$ and $p$, we set $N = c$, $p = 1$, and the likelihood is maximized at 1. But these

231

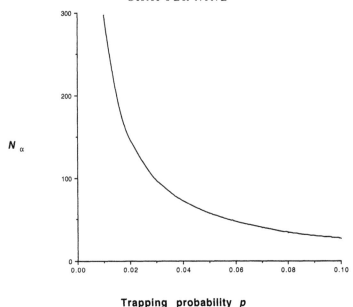

**Trapping probability *p***

FIGURE 9.4. The value of $N_\alpha$, showing how Bayesian analysis can be used to estimate the number of animals in a region, given no trap catch.

are surely values that no reasonable person would choose! There are three ways around this. First, one could estimate $p$ from other data, using search theory (Mangel and Beder 1985). Second, one could modify the entire operation; this would work if one initially tagged animals and then conducted resightings. Third, we could adopt a fully Bayesian approach in which we choose a prior for $p$ as well as a prior for $N$. The likelihood now explicitly depends on $N$ and $p$, so we write

$$\mathcal{L}\{c\,|\,N,p\} = \binom{N}{c} p^c (1 - p)^{N-c}$$

for $N = c, c + 1, c + 2, \ldots,$ and $0 \le p \le 1$. (9.44)

We could then, for example, pick the uniform prior for $N$ and the prior that is proportional to $p^A(1 - p)^B$ for $p$:

$$f_{\text{post}}(N,p|c) = \frac{\binom{N}{c} p^c (1-p)^{N-c} p^A (1-p)^B}{\sum_{N'=c}^{\infty} \int_{p'=0}^{1} \binom{N'}{c} (p')^c (1-p')^{N'-c} (p')^A (1-p')^B dp'}, \tag{9.45}$$

which simplifies to

$$f_{\text{post}}(N,p|c) = \frac{\binom{N}{c} p^{c+A} (1-p)^{N+B-c}}{\sum_{N'=c}^{\infty} \binom{N'}{c} \int_{p'=0}^{1} (p')^{c+A} (1-p')^{N'+B-c} dp'}, \tag{9.46}$$

and provides a "doubly" Bayesian method when both N and p are unknown.

## MODEL VERSUS MODEL VERSUS MODEL

We can use Bayesian analysis in a more general setting to consider not only distributions of parameters, but posterior distributions of models. This is another advantage of the Bayesian method.

Given the data, we fit the parameters of each model, say by maximum likelihood, and we denote the maximum value of the likelihood of the data given model $M_i$ by $\mathcal{L}^*\{\text{data}|M_i\}$. The Bayesian approach naturally allows us to consider the posterior probability of model $i$, given the observed data, since

$$\Pr\{M_i \text{ given the data}\} = \frac{\Pr\{M_i \text{ and the data}\}}{\Pr\{\text{data}\}}. \tag{9.47}$$

If $\Pr\{M_i\}$ is the prior probability of model $i$, then

$$\Pr\{M_i \text{ given the data}\} = \frac{\Pr\{\text{data given } M_i\} \Pr\{M_i\}}{\Pr\{\text{data}\}}$$

$$= \frac{\mathcal{L}^*\{\text{data}|M_i\}\Pr\{M_i\}}{\sum_j \mathcal{L}^*\{\text{data}|M_j\}\Pr\{M_j\}}. \tag{9.48}$$

233

If we assume that each model is a priori equally likely, then Equation 9.48 simplifies to

$$\Pr\{M_i \text{ given the data}\} = \frac{\mathscr{L}*\{\text{data}|M_i\}}{\sum_j \mathscr{L}*\{\text{data}|M_j\}}. \tag{9.49}$$

Alternatively, we might suppose that there is an a priori probability $p$ that model 1 is correct, and that all the other models (say there are $M$ models) are equally likely, with probability $(1 - p)/(M - 1)$. The Bayesian approach then leads to the following conclusion about the posterior probability of model 1, given the data

$$\Pr\{M_1 \text{ given the data}\}$$
$$= \frac{\mathscr{L}*\{\text{data}|M_1\}p}{\mathscr{L}*\{\text{data}|M_1\}p + (1 - p)/(M - 1) \sum_{j=2}^{M} \mathscr{L}*\{\text{data}|M_j\}}. \tag{9.50}$$

We can then, for example, make a plot of $\Pr\{M_1$ given the data$\}$ versus $p$ to see how the data have shifted the prior belief about the likelihood of model 1. We shall do exactly this in the next chapter.

# Management of Hake Fisheries in Namibia

## MOTIVATION

Quantitative methods have a long history in fisheries science (Smith 1994), because fisheries scientists recognized early on that their problems are in many ways much more difficult than terrestrial ones. For example, it is difficult to estimate abundance when one cannot see the population. Perhaps the major impetus was the need to set regulations; this has driven the collection and analysis of data. In some fisheries, such as those for Pacific salmon in the United States and Canada, data are collected and analyzed and regulations are set on a daily basis. Most fisheries involve large-scale perturbations of ecological systems, systematic data collection, and pressing financial and political needs for scientific advice. Because fisheries management is usually a public policy decision and most fisheries retain some form of common access, there can be considerable public scrutiny of fisheries decisions.

The ecological questions asked in the area of fisheries management range widely, from the definition of species through aquatic toxicology, dynamics of lakes and marine ecosystems, to the population dynamics of exploited fish stocks. Our experience is primarily with the last category, particularly focusing on how fish stocks have responded to exploitation, which is what we consider in this chapter. The broad issue of harvesting usually involves simultaneous production of catch and conservation of the stock for the purposes of future catch. Thus we ask questions such as:

235

- Would long-term yields be improved if catch decreased temporarily or permanently?
- Are the current levels of yields sustainable?
- What is the long-term potential yield?
- Is the stock in danger of collapse?
- How large is the population now, and how large was it when the fishery began?

None of these questions fits into the classic Popperian battle between a single model and the data. Rather, these questions always involve the competition between hypotheses about the dynamics of the stock and the interactions between the stock, the ecosystem, and management. The job of the ecological detective is to provide the best possible scientific information on which decisions can be based. Thus, we want to understand the relative likelihood of different possible states of the fish stock, and of how the stock might respond to different management actions.

For simplicity in our analysis, we focus to a great extent on the distribution of maximum sustainable yield (MSY). Although MSY has been pretty much discredited as a management objective for many years (e.g., Clark 1985), it is a useful pedagogic tool because often we can compute MSY easily from models of the population dynamics. It is also convenient for illustration of the difference between those cases in which MSY can be viewed as a direct parameter of the model and those cases in which it cannot.

In this chapter, we illustrate the use of the AIC to select between non-nested models, how Bayesian methods can be used to incorporate knowledge gained from other studies and to understand uncertainty in MSY, and how models that are biologically better may be statistically poorer. This final point constantly arises in applied problems.

## THE IMPACT OF ENVIRONMENTAL CHANGE

Most models of fish population dynamics ignore environmental change except as a form of "white noise" that affects the recruitment to the stock. However, there is a growing body of evidence that a considerable component of the changes seen in many fish stocks has been due to environmental changes (Caddy and Gulland 1983; Hilborn and Walters 1992). What could be viewed as overfishing may, in reality, be a natural decline due to environmental changes.

The problem is that on the time scale that data are usually available, it is difficult (if not impossible) to distinguish between a change in stock abundance due to systematic environmental changes and a change in stock abundance due to fishing pressure (but also see Hutchings and Meyers 1994). Since fishing pressure can be managed but the environment cannot, the default assumption in fisheries models and management has been to assume that the changes are due to fishing pressure. Thus, we use models without systematic environmental change and leave the challenge of realistically considering environmental change for the next generation of ecological detectives.

THE ECOLOGICAL SETTING

The Namibian fishery for two species of hake (*Merluccius capensis* and *M. paradoxus*) was managed by the International Commission for Southeast Atlantic Fisheries (ICSEAF) from the mid-1970s until about 1990. Our analysis will be concerned with the period up to and including ICSEAF management. Hake were fished by large ocean-going trawlers primarily from Spain, South Africa, and the Soviet Union. While both species are captured in the fishery, the fishermen are unable to distinguish between them, and both are treated as a single stock for management purposes. Here we focus on ICSEAF statistical regions 1.3 and 1.4 (Figure 10.1a). As the fishery developed, essentially without any regulation or conservation organization, the catch per unit effort (CPUE), measured in tons of fish caught per hour, declined dramatically until concern was expressed by all fishing nations. The profitability of fishing depends primar-

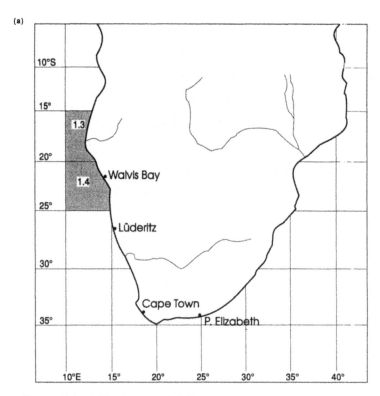

FIGURE 10.1. (a) The location of ICSEAF regions 1.3 and 1.4.

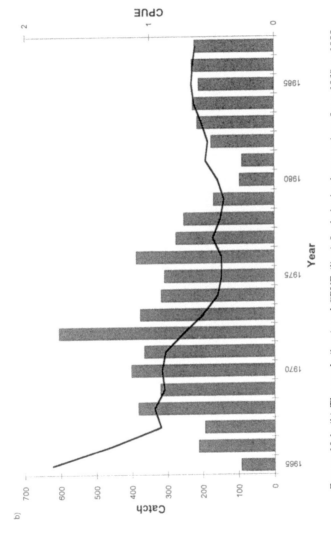

FIGURE 10.1. (b) The catch (bars) and CPUE (line) for hake in these regions from 1965 to 1988.

TABLE 10.1.  Catch and CPUE data for Namibian hake.

| Year | CPUE (tons per standardized trawler hour) | Catch (thousands of tons) |
|---|---|---|
| 1965 | 1.78 | 94 |
| 1966 | 1.31 | 212 |
| 1967 | 0.91 | 195 |
| 1968 | 0.96 | 383 |
| 1969 | 0.88 | 320 |
| 1970 | 0.90 | 402 |
| 1971 | 0.87 | 366 |
| 1972 | 0.72 | 606 |
| 1973 | 0.57 | 378 |
| 1974 | 0.45 | 319 |
| 1975 | 0.42 | 309 |
| 1976 | 0.42 | 389 |
| 1977 | 0.49 | 277 |
| 1978 | 0.43 | 254 |
| 1979 | 0.40 | 170 |
| 1980 | 0.45 | 97 |
| 1981 | 0.55 | 91 |
| 1982 | 0.53 | 177 |
| 1983 | 0.58 | 216 |
| 1984 | 0.64 | 229 |
| 1985 | 0.66 | 211 |
| 1986 | 0.65 | 231 |
| 1987 | 0.63 | 223 |

ily on the CPUE, so that if the CPUE declines, profits will also decline. The concern about the dropping CPUE led to the formation of ICSEAF and subsequent reductions in catch. After catches were reduced, the CPUE began to increase (Table 10.1); also see Punt (1988). In the data used in this analysis, the CPUE is the catch per hour of a specific class of Spanish trawlers. Such a definition is used to avoid bias due to increasing gear efficiency or differences in fishing pattern by different classes or nationalities of vessels.

## THE DATA

We commonly find two types of data in fisheries harvesting problems. The first is a history of catches removed from the stock (Figure 10.1b). The second is an index of abundance; some measure that indicates the size of the stock. Other information (e.g., knowledge of the age structure of the population, individual growth rates, fecundity at different ages, breeding seasons, or other basic biology) is almost always available and may be very useful. It is often extremely important to know if the stock was unfished at the beginning of the data series—if this is the case, the problem in estimation of parameters is greatly simplified.

In addition, we often know something about the experience of fisheries for the same or similar species in other locations in the world; this kind of information can be incorporated by using Bayesian analysis. For instance, herring, anchovy, and sardine exhibit intense schooling behavior. Thus, when fish are caught they are usually part of a large school and the CPUE is high, which makes it a very poor index of abundance, because even when total abundance is low the fish re-form to a few high-density schools. Such stocks have frequently collapsed under heavy fishing pressure even though the CPUE remained high. Hake and their relatives do not school so intensely, and it is generally believed that the CPUE is a better index of abundance for such species.

We also may know how quickly different taxonomic or life history groups of fish have recovered when fishing pressure is reduced. We may have information about the sensitivity of recruitment of juveniles to the total spawning abundance— marine mammals and sharks with low fecundity are especially sensitive to the size of their spawning stock. On the other hand, some groups of fish, such as cod and hake, have proved to be remarkably resilient to reduced spawning biomass. In these species, recruitment appears to depend weakly on the size of the spawning stock. The compilation and use of information from other stocks is often called "meta-analysis" (Fernandez-Duque and Valeggia 1994)

We thus classify data into four broad categories: (1) exploitation history of the stock, (2) basic biology of the species, (3) history of exploitation on similar stocks elsewhere, and (4) knowledge of the mechanics of the fishing and data collection processes.


## THE MODELS

We explore two models of fish population dynamics. The first is the Schaefer model, which is based on logistic population dynamics. The second is a more elaborate model that explicitly deals with life history phenomena, such as age of recruitment, survival, and growth. Both models use a single variable, stock biomass, to represent the abundance of the stock. Thus, we call them "biomass dynamics" models, since the focus of the model is on the dynamics of the vulnerable stock biomass, although they are more commonly referred to as "surplus production models" in the fisheries literature. We do not consider age-structured models, although such models are used for a significant proportion of the world's fish stock assessments. We encourage the interested reader to extend our ideas into this domain. Hilborn and Walters (1992) give a general introduction.


### The Schaefer Model

The Schaefer model appends a catch $C_t$ to a standard logistic model for the biomass dynamics. Thus, if $B_t$ is the biomass of the stock that is vulnerable to fishing at the start of period $t$, we assume that

$$B_{t+1} = B_t + rB_t \left( 1 - \frac{B_t}{K} \right) - C_t, \tag{10.1}$$

where $r$ is the growth rate, $K$ is the equilibrium size of the population in the absence of catch, and $C_t$ is the catch. We often assume that

$$C_t = q_0 E_t B_t, \qquad (10.2)$$

where $q_0$ is the "catchability coefficient" and $E_t$ is the "fishing effort" during period $t$.

The index of abundance $I_t$ is generally assumed to be proportional to biomass,

$$I_t = q B_t. \qquad (10.3)$$

There are a number of standard measures of performance derived from this simple model (Clark 1990; Krebs 1994). These include the per capita harvest rate for maximum sustained yield

$$h_{MSY} = \frac{r}{2}, \qquad (10.4)$$

which is often called the "optimal harvest rate." The harvest rate is the fraction of the stock that is removed by harvesting. The stock size at MSY is

$$B_{MSY} = \frac{K}{2}. \qquad (10.5)$$

Combining Equations 10.4 and 10.5 gives the MSY:

$$MSY = \frac{rK}{4}. \qquad (10.6)$$

The MSY is found by assuming that $B_t = B_{t+1}$ (assuming a steady state) and solving for the biomass level at which a constant harvest is maximized. Finally, the virgin (unfished) biomass is

$$B_{virgin} = K. \qquad (10.7)$$

We convert the deterministic model in Equations 10.1–10.3 to a stochastic one by adding process and observation uncertainty:

$$B_{t+1} = \left[ B_t + rB_t \left( 1 - \frac{B_t}{K} \right) - C_t \right] \exp \left( W_t \sigma_W - \frac{\sigma_W^2}{2} \right),$$

$$I_t = qB_t \exp \left( V_t \sigma_V - \frac{\sigma_V^2}{2} \right), \tag{10.8}$$

where $W_t$ and $V_t$ represent process and observation uncertainty, respectively, and are normally distributed random variables with mean 0 and standard deviations $\sigma_W$ and $\sigma_V$, respectively. It is common practice to assume log-normal distributions for the observation and process errors because (1) the random processes are usually multiplicative, and (2) using a normal distribution could lead to negative values of biomass or index of abundance (compare this with Chapter 8).

*A Model with Lagged Recruitment, Survival, and Growth (LRSG)*

The logistic model does not explicitly deal with growth, recruitment, or survival and does not incorporate lags to recruitment. In the logistic model, the relationship between net growth or recruitment (also known as surplus production) and stock size is fixed in that the biomass that produces the MSY is always $K/2$. In addition, the logistic model allows only a one-year time lag between changes in biomass and changes in net production.

A more flexible model that incorporates alternative life history characteristics is the "lagged recruitment, survival, and growth" (LRSG) model. This model is a simple approximation to the delay-difference model of Deriso (1980). First, biomass in any year is the balance of the surviving biomass from the previous year, recruitment, and catch, so that

$$B_{t+1} = sB_t + R_t - C_t. \tag{10.9}$$

If the model dealt with numbers of individuals rather than biomass, then $s$ would represent survival from all causes except for fishing from one year to the next. When we focus on biomass and recognize that individuals grow in mass, then $s$ reflects how much biomass changes from year to year due to both survival and growth. For example, if the survival from one year to the next is 80% and surviving individuals grow about 10% in mass each year, then $s = (0.8) \times (1.1) = 0.88$.

$R_t$ is the recruitment to the population (the addition of biomass that is now vulnerable to fishing), which we assume to be

$$R_t = \frac{B_{t-L}}{a + B_{t-L}},$$

$$R_0 = B_0(1 - s). \tag{10.10}$$

Here, $B_0$ is the virgin biomass (analogous to $K$ in the Schaefer model), and the index $t - L$ indicates that recruitment in year $t$ depends on biomass $L$ years before (hence the word "lag" as a description of this model); $L$ represents the number of years from egg deposition until the fish are vulnerable to the fishing gear. In most fisheries, the transition from being young, small, and not vulnerable to fishing to being old, large, and vulnerable to fishing is gradual, but in this model we assume what is commonly known as "knife-edge" selectivity: in one year the fish are not vulnerable to the gear and the next year they are. Further, we assume that individuals become both vulnerable to the fishery and reproductively mature at the same age (alternatives are described by Mangel [1992]).

The parameters $a$ and $b$ are defined by

$$a = \frac{B_0}{R_0}\left(1 - \frac{z - 0.2}{0.8z}\right),$$

$$b = \frac{z - 0.2}{0.8zR_0}, \tag{10.11}$$

where the parameter $z$ scales the sensitivity of recruitment to biomass at the time of spawning. Recruitment described by Equations 10.10 and 10.11 is called a "Beverton-Holt stock

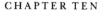

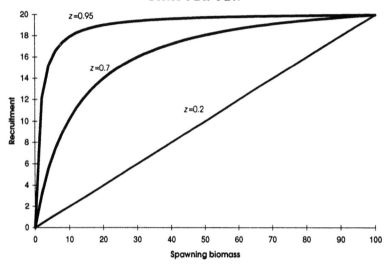

FIGURE 10.2. Beverton-Holt stock recruitment curves with different z values.

recruitment curve" (Beverton and Holt 1993); see Figure 10.2. The parameters $R_0$, $a$, and $b$ are derived quantities (Hilborn and Walters 1992, 88 ff.). The key characteristic of the Beverton-Holt recruitment curve is that $R_t$ never decreases with increasing spawning stock. The initial slope of the curve is $1/a$, and the asymptote is $1/b$. The parameter $R_0$ is the recruitment when spawning biomass is $B_0$, and the parameter $z$ (the "steepness") represents how steeply the curve ascends, and is, by definition, the ratio between recruitment at $0.2B_0$ and $R_0$ (Figure 10.2). Thus, if $z = 0.99$, recruitment is almost constant; if $z = 0.2$, recruitment is proportional to spawning stock; and if $z = 0.7$, then at $0.2B_0$ recruitment is 70% of what it was at $B_0$. We prefer to use the parameters $z$ and $B_0$ instead of $a$ and $b$ because $z$ and $B_0$ have straightforward biological interpretations. In addition, one can obtain prior probability distributions for $z$ by analysis of other fish stocks with similar biology.

As before, we assume an index of abundance $I_t$ proportional to biomass:

$$I_t = qB_t. \tag{10.12}$$

This model has five parameters: survival $s$, the time lag $L$ between reproduction and recruitment, the recruitment parameter $z$, the unfished biomass $B_0$, and the scaling factor $q$ relating biomass to the index of abundance.

The MSY and the biomass at MSY, $B_{MSY}$, are computed by setting $B_t = B_{t+1}$ and then maximizing catch with respect to biomass. The steady-state catch is

$$C = B(s - 1) + \frac{B}{a + bB} \tag{10.13}$$

so that the catch is maximized by setting

$$\frac{dC}{dB} = s - 1 + \frac{1}{a + bB} - \frac{bB}{(a + bB)^2} = 0, \tag{10.14}$$

from which we find the biomass at MSY to be

$$B_{MSY} = \frac{1}{b}\sqrt{\frac{a}{1 - s}} - a. \tag{10.15}$$

Substituting $B_{MSY}$ into the steady-state–catch relationship Equation 10.13 gives

$$MSY = B_{MSY}\left(s - 1 + \frac{1}{a + bB_{MSY}}\right). \tag{10.16}$$

The biomass at the MSY and the MSY are not simple functions of the life history parameters, but they are not intractable either. Equations 10.10–10.16 become a stochastic model in a manner similar to that used in Equation 10.8; we encourage you to do this before reading on.

## THE CONFRONTATION

### Schaefer Model with Observation Uncertainty

As described in Chapter 7, to achieve relative tractability of computation, we must assume either process uncertainty or observation uncertainty, but not both. For the Schaefer model with observation uncertainty, we compare predicted and observed values of the CPUE, since the CPUE is the index of abundance. The principal data will be the history of catches, and the parameters that must be estimated are the stock biomass $B_0$ at the beginning of the data series, the intrinsic rate of growth $r$, the carrying capacity $K$, and the catchability coefficient $q$. Because we assume only observation uncertainty, the stock dynamics are deterministic. We also use the simplifying assumption that the stock was unfished at the beginning of the data series, so that $B_0 = K$. This reduces the parameters to be estimated to $r$, $K$, and $q$. The equations for the predicted index of abundance ($I_{est,t}$) are

$$B_{est,t+1} = B_{est,t} + rB_{est,t}\left(1 - \frac{1}{K}B_{est,t}\right) - C_t,$$

$$B_{est,0} = K,$$

$$I_{est,t} = qB_{est,t}. \tag{10.17}$$

Given values of $r$, $K$, and $q$ and the history of catches, these equations allow us to predict the index of abundance, which is compared to the observed index $I_t$. Since the index of abundance is assumed to have a log-normal distribution, the negative log-likelihood in a single period is

$$\mathbf{L}_t = \log(\sigma_V) + \frac{1}{2}\log(2\pi) + \frac{[\log(I_{est,t}) - \log(I_t)]^2}{2\sigma_V^2}. \tag{10.18}$$

The total negative log-likelihood is the sum over $t$ of all the individual negative log-likelihoods and is minimized across the parameters $r$, $K$, $q$, and $\sigma_V$.

---

Pseudocode 10.1

1. Input the catch and CPUE data.
2. Input starting estimates of the parameters $r$, $K$, $q$, and $\sigma_V$.

3. Find values of the parameters that minimize the total negative log-likelihood through the following steps:
   (a) Predict values of $B_{est}$ and $I_{est}$ from Equation 10.17.
   (b) Calculate the negative log-likelihood using Equation 10.18.
   (c) Sum the negative log-likelihoods over all years.
   (d) Minimize the total negative log-likelihood.

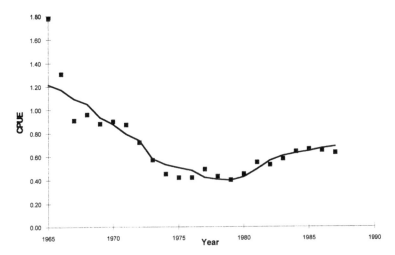

FIGURE 10.3. Observed index of abundance (squares) for the hake fishery and best fit (line) for the Schaefer model with observation uncertainty.

As in Chapter 8, all parameters and biomasses must be constrained to be non-negative.

Employing this pseudocode, the estimated values of the parameters are $r = 0.39$, $K = 2709$, $q = 0.000\ 45$, and $\sigma_V = 0.12$ (Figure 10.3), so that the MSY $= 266$. The next step is to understand the level of certainty in the two biologically important parameters of interest ($r$ and $K$) and through them the level of certainty in the MSY. This can be achieved by systematically searching over $r$ or $K$ and finding the values of the other parameters that minimize the negative log-likelihood. There is some help here. Punt (1988) derived an analytic solution for the estimate of $q$. Given $r$, $K$, and $n$ time periods, the estimate of $q$ that minimizes the negative log-likelihood is

$$\hat{q} = \exp\left(\frac{1}{n}\sum_{t=1}^{n}\left[\log(I_t) - \log(B_{est,t})\right]\right).$$

(10.19)

Note that $r$ and $K$ appear in this expression implicitly through the value of estimated biomass, and that the expression is independent of $\sigma_v$. A pseudocode for the likelihood profile for $r$ is:

---

Pseudocode 10.2

1. Input the catch and CPUE data.
2. Input starting estimates of $K$ and $\sigma_V$ and the desired ranges and step sizes for $r$.
3. Systematically loop over values of $r$.
4. For each value of $r$, find the values of $K$ and $\sigma_V$ that minimize the negative log-likelihood, as previously done, except that $r$ is fixed at a specific value in step 3 and Equation 10.19 is used.
5. For each value of $r$, calculate the value of the $\chi^2$ distribution.

---

The calculation of the $\chi^2$ value is based on the fact that twice the difference between the negative log-likelihood for any value of $r$, $\mathbf{L}(r)$, and the lowest value of negative log-likelihood obtained, $\mathbf{L}_{min}(r)$, is $\chi^2$ distributed with one degree of freedom.

Employing this pseudocode (Figure 10.4a), we find that the 95% confidence bounds on $r$ are roughly from 0.325 to 0.475. The likelihood profile for $K$ (Figure 10.4b) is obtained similarly.

250

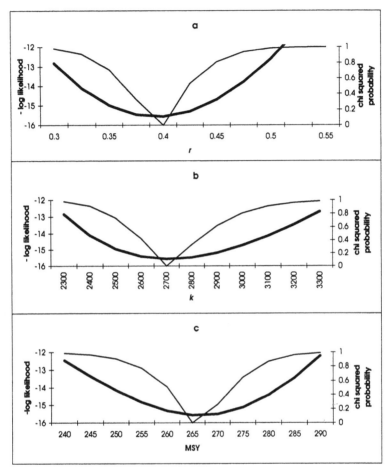

FIGURE 10.4. The likelihood profile and chi-square probability associated with the profile for $r$ (a), $K$ (b), and MSY (c) for the Schaefer model assuming observation uncertainty. The thick line is the negative log-likelihood, the thin line is the chi-square probability.

We are really interested in the likelihood profile for the MSY. The easiest way to profile the MSY is to recognize that since MSY $= rK/4$, we can redefine the parameters of the model as $r$ and MSY, in which case $K = 4MSY/r$. The likelihood profile for the MSY is shown in Figure 10.4c.

*Schaefer Model with Process Uncertainty*

If we assume process uncertainty, the predicted biomass depends on the observed index of abundance rather than on the predicted biomass in the previous year. If observations are perfect, the true biomass is $I_t/q$. The equations underlying the analysis are

$$B_{est,t+1} = \frac{1}{q} I_t + r \frac{1}{q} I_t \left( 1 - \frac{1}{Kq} I_t \right) - C_t,$$

$$I_{est,t} = q B_{est,t}. \tag{10.20}$$

As before, we compare $I_{est,t}$ with the observed index $I_t$ so that the negative log-likelihood in a single period is

$$\mathbf{L}_t = \log(\sigma_W) + \frac{1}{2} \log(2\pi)$$

$$+ \frac{[\log(I_{est,t}) - \log(I_t)]^2}{2\sigma_W^2}. \tag{10.21}$$

There are some computational differences in a confrontation based on observation uncertainty and one based on process uncertainty (see Chapter 7). First, we change the prediction equation so that $B_{est,t+1}$ depends on $I_t$ instead of on $B_{est,t}$. Second, we no longer require an estimate of initial biomass. Third, the only usable observations are consecutive ones. If we do not know the index of abundance in consecutive time periods, we cannot predict $I_{t+1}$ from $I_t$, and this approach cannot be used. In addition, there is no longer an analytic form for the estimate of $q$, so we must estimate $r$, $K$, and $q$ (Table 10.2).

There are no major differences in the MSY estimated (266 versus 278). However, there are major differences in the confidence bounds on the parameters. To see this, we compare the likelihood profile for the MSY (Figure 10.5) to

TABLE 10.2.  Estimated parameters with the Schaefer model.

| Parameter | Observation uncertainty | Process uncertainty |
|---|---|---|
| $r$ | 0.39 | 0.32 |
| $K$ | 2709 | 3519 |
| $\sigma_V$ or $\sigma_W$ | 0.12 | 0.10 |
| MSY | 266 | 278 |
| $q$ | 0.000 45 | 0.000 26 |

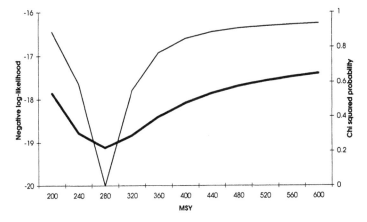

FIGURE 10.5. The likelihood profile for MSY from the Schaefer model assuming process uncertainty (compare to Figure 10.4c).

Figure 10.4c: assuming observation uncertainty leads to a narrower confidence region.

### LRSG Model

For the LRSG model, Equations 10.9–10.16, which explicitly includes survival, recruitment, and a time lag to recruitment, the likelihood calculations are similar to those for the Schaefer model, but we use a different model of population dynamics. We only consider results with observation uncertainty.

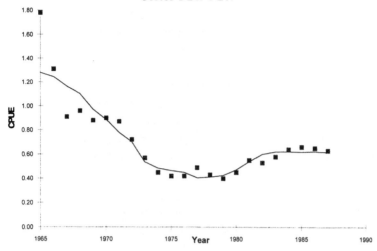

FIGURE 10.6. Observed index of abundance (squares) for the hake fishery and best fit (line) for the LRSG model with observation uncertainty.

Based on the biology of hake, we set $L = 4$ and find that the parameter values that minimize the negative log-likelihood (Figure 10.6) are $B_0 = 3216$, $s = 0.87$, $z = 0.99$, and $q = 0.000\ 40$. The negative log-likelihood is $-16.88$, which is slightly better than that obtained from the Schaefer model ($-15.56$). The values of the parameters are consistent with the biology of hake. For example, natural mortality is about 20% per year, and there is roughly 10% growth in body mass per year, so that the value of $s$ should be on the order of 0.90. The value $z = 0.99$ indicates that recruitment is nearly constant, which is also consistent with current knowledge of the biology of hake. Finally, $B_0 = 3216$ is consistent with the catches that have been removed from the stock.

Calculating the likelihood profile of the MSY in the LRSG model is more difficult, because there is no simple relationship between any individual parameter and the MSY (Equation 10.16). To obtain the best fit for a fixed MSY, we add a penalty function to the likelihood, thus forcing the values of

254

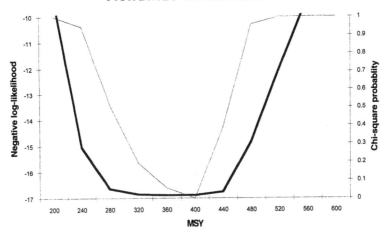

FIGURE 10.7. Likelihood profile of the MSY from the LRSG model. The thick line is the negative log-likelihood, the thin line is the chi-square probability.

$B_0$, $s$, and $z$ to make the MSY very close to the target MSY, and minimize

$$F(B_0,s,z) = \sum_t \mathbf{L}_t + c_p\{\mathrm{MSY}(B_0,s,z) - \mathrm{MSY}_{\mathrm{profile}}\}^2, \quad (10.22)$$

where $\mathbf{L}_t$ is the negative log-likelihood for year $t$, $c_p$ is the penalty cost (chosen so that deviations from the target MSY are of the same magnitude as the negative log-likelihood), $\mathrm{MSY}(B_0,s,z)$ is the value of the MSY for the specific value of the parameters, and $\mathrm{MSY}_{\mathrm{profile}}$ is the target value of the MSY in the profile. By minimizing $F(B_0,s,z)$, we find the best set of parameters that are consistent with an MSY equal to $\mathrm{MSY}_{\mathrm{profile}}$ (Figure 10.7).

The key difference in results between the LRSG model and the Schaefer model is that the LRSG model admits much more uncertainty in the MSY. We cannot use the likelihood ratio test to compare the models, because the LRSG model is not nested with the Schaefer model, but we can use the Akaike information criterion (AIC) as a guide. The LRSG model has six parameters: $B_0$, $s$, $z$, $\sigma_V$, $q$, and $L$. The Schaefer model has four parameters: $K$, $r$, $\sigma_V$, and $q$. If we assume the lag to recruitment is known, the LRSG has one more free parameter than the Schaefer model. Thus, the negative log-likelihood would need to be approximately two less than the negative log-likelihood of the Schaefer model. The negative log-likelihoods were $-16.88$ (LRSG) and $-15.56$ (Schaefer). Using the AIC, we conclude that the Schaefer model is a better choice. That is, the AIC indicates that the LRSG model does not provide a significant improvement in fit over the Schaefer model, taking the number of parameters into account. If the question is which model best represents the uncertainty in the MSY, the comparison provides less guidance. The tight confidence bounds from the Schaefer model are due primarily to its very specific structural assumptions about population dynamics. The LRSG model, on the other hand, has much more flexibility in the description of the biology. This is a case in which a simpler, but certainly less biologically accurate, description wins the statistical confrontation, in part because of limited data, and in part because the more complicated model allows a priori a wider range of biological dynamics and is penalized by the AIC because of this. However, if the task is to make the best appraisal of uncertainty in the MSY, we should attempt to incorporate all known information about the species. This requires a Bayesian approach.

BAYESIAN ANALYSIS OF THE LRSG PARAMETERS

A Bayesian approach allows us to specify prior distributions for the parameters of the LRSG model, using knowledge about the biology of the hake. For example, almost all hake and related species show very little reduction in recruitment with reductions in spawning biomass, so the steepness is likely close to 1. We know from biological studies that the survival is roughly 80% per year and that increase in mass per year is about 10%.

Thus, to compute a better description of the uncertainty in the MSY, we conduct a Bayesian analysis in which priors specify prior knowledge about $s$ and $z$. The Bayesian analysis requires integrating over the five parameters and specifying a prior for each one. As a shortcut we use the analytic formulas available for the values of $\sigma_V$ and $q$ that maximize the likelihood, and only integrate over $B_0$, $s$, and $z$. This reduces admitted uncertainty but is a useful computational shortcut.

We use a Monte Carlo form of integration, in which we draw a random value of each parameter from its prior distribution, then calculate the likelihood for this combination of parameters. Repeating this process 10 000 times approximates integration over a specific range of the values of the parameters. Because we are relatively confident about the ranges of $B_0$, $s$, and $z$ but less certain about their distributions, we use uniform prior distributions. Thus, $B_0$ is uniformly distributed from 0 to 7000, $s$ is uniformly distributed from 0.65 to 0.95, and $z$ is uniformly distributed from 0.8 to 1.0. If an analysis of data from other hake-like species were available, we might use a more informative distribution for the prior for $z$, and existing age-structure information could be used to formulate a prior for $s$. For example, McAllister et al. (1994) used several historical data sets to formulate prior distributions for hoki, another hake-like species.

A pseudocode for the Monte Carlo–Bayesian integration of the LRSG model is:

---

Pseudocode 10.3

1. Input the catch and CPUE data.
2. Input low and high values for $B_0$, $s$, and $z$.
3. Randomly draw values of $B_0$, $s$, and $z$ from their prior distributions.
4. Project the stock biomass forward using these parameters, using Equations 10.9–10.11.

5. Calculate the values of $q$ and $\sigma_V$ that maximize the likelihood.
6. Calculate the MSY associated with the parameters.
7. Repeat steps 3–6 ten thousand times.
8. Divide the outputs of interest ($B_0$, $s$, $z$, and MSY) into discrete intervals and calculate the proportion of the total likelihood that falls within each interval. Make sure you use total likelihood and not negative log-likelihood.

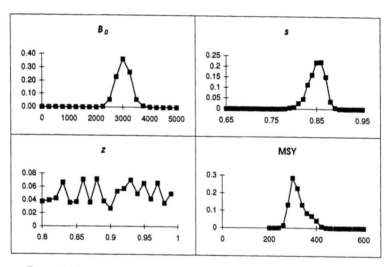

FIGURE 10.8. Bayes posteriors for $B_0$, $s$, $z$, and MSY from the LRSG model.

Doing this (Figure 10.8) shows that $B_0$, $s$, and the MSY are well but that $z$ is poorly characterized. These graphs are the Bayes posterior distributions of the parameters given the priors, the data, and the model. We now compare (Figure 10.9) the distributions of the MSY estimated from the likelihood profiles of the Schaefer model, the likelihood profile of the LRSG model, and the Bayes posterior of the LRSG model. The likelihood profiles have been rescaled so that the area under the curve is equal to the area under the Bayes posterior. We see the very tight distribution of the

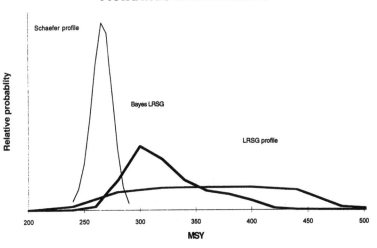

FIGURE 10.9. The normalized likelihood profile for the MSY from the Schaefer and LRSG models, and the Bayes posterior for the MSY from the LRSG model.

Schaefer model, the very broad distribution of the likelihood profile of the LRSG model, and the Bayes posterior of the LRSG model. Were we asked which of these best represented our state of understanding of the MSY, we would opt for the Bayes posterior, because it incorporates more biological understanding than the other two models.

As with all Bayesian analysis, the results may depend on the prior distributions, and the first check is to see what influence the priors may have had on the results. The simplest way to do this is to repeat the previous calculations, but setting the likelihood for every set of parameters equal to 1. We are in effect asking, "What are the posteriors if the data tell us nothing?" The results of uniform priors for $B_0$, $s$, and $z$ must be uniform posteriors on these three parameters (Figure 10.10). However, the assumed priors also make an MSY of 600 or greater by far the most likely. By comparing Figures 10.8 and 10.10, we see that the data provided a lot of information about $B_0$ and $s$, but essentially no information about $z$. Most importantly, the well-defined distributions

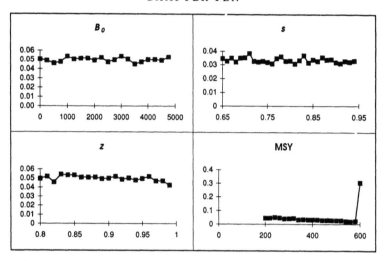

FIGURE 10.10. Posterior distribution of the parameters from the LRSG model when all likelihoods are set to 1. This shows the posteriors implied by the priors. The peak for the MSY at 600 indicates that many combinations of priors yield an MSY of 600 or greater.

(particularly for the MSY) we found from the full Bayesian analysis did not come from the priors.

## IMPLICATIONS

In summary, we compared the simpler Schaefer model with four parameters ($r$, $K$, $q$ and $\sigma_V$) to the more complex LRSG model with five parameters ($B_0$, s, z, $q$, and $\sigma_V$). We failed to obtain a significantly better fit according to the AIC, and based on this conclude that the Schaefer model "won the confrontation."

The general approach is that the "best" model is the one that is most consistent with the data and with the fewest free parameters. This is the basis of the likelihood ratio test or the Akaike information criterion. If the goal were simply to fit the data, then we could stop now. But management goals often involve making decisions based on the modeling. In that case, when we have more information than very narrowly defined data (which in this case were the catch and the CPUE), we should take advantage of it. Although the catch and CPUE data do not allow us to say that the LRSG model is more reasonable than the Schaefer model, our biological knowledge does. The reason for the relatively tight confidence bounds on the MSY in the Schaefer model is that the structure of the model is very specific. In the Schaefer model, net production declines once stock falls below $0.5B_0$, while the LRSG model allows much more flexibility in the relation between net production and stock size.

The Bayesian approach offers an alternative method of model selection. Imagine that we assigned the Schaefer and LRSG models equal prior probability. Since the likelihoods are almost the same (suggesting that the data are not helpful in separating the two models), the Bayesian posteriors for the two models will be roughly equal. However, it is not appropriate to assign equal prior probability to the Schaefer and LRSG models. The behavior of the Schaefer model can be mimicked by choosing the parameters in the LRSG model carefully, but the reverse is generally not true. Thus, one might argue that a priori the LRSG model is far more likely than the Schaefer model.

This confrontation is an example of the general question of when to stop adding more complexity to models. If we abandon the traditional criteria, such as likelihood ratio and AIC, we can admit more and more complex models. If the purpose is prediction, this is not appropriate because parsimony is desirable, but if we want to understand the true uncertainty, it pays to consider the broadest range of possible models.

Thompson (1992) provides another example of a Bayesian approach to management advice when stock-recruitment parameters are uncertain. Walters and Punt (1994) show how Bayesian analysis can be used to describe the posterior probability of sustainable catch for a fishery in which

261

virtual population analysis (VPA) and survey data are used to estimate current stock size and allowable catch. Such Bayesian information can form the basis for risk analysis of fisheries (Cordue and Francis 1994). Speed (1993) and Schnute (1993) provide other examples of how the techniques of likelihood and the likelihood profile (among others) can be used in the analysis of fisheries problems. The most intensive Bayesian analysis of a fisheries model is by McAllister et al. (1994).

The Namibian hake fishery illustrates several of the common problems faced by the ecological detective who works on applied problems. The data are not determined in controlled experiments and involve a completely untested assumption (that the CPUE is proportional to abundance). As a result, the answers that we obtain are not clean or clear. Usually, examples given in textbooks are the ones in which we can answer questions and estimate parameters with considerable confidence, but in natural resource management the uncertainties are often much larger.

This example also illustrates common problems in model selection and in uncertainty. The likelihood ratio or AIC methods would choose the Schaefer model and thus underrepresent the uncertainty in the MSY. If we estimate the uncertainty in the MSY using the LRSG model alone, we fail to incorporate considerable prior knowledge about the parameters. The Bayesian approach lets us incorporate knowledge of the biological parameters and produce a better estimate of the true uncertainty. It also highlights the importance of distilling historical knowledge into current assessments of alternative hypotheses. If we fail to use what we have learned in previous studies, we will learn very slowly indeed.

# The Confrontation: Understanding How the Best Fit Is Found

## INTRODUCTION

In this chapter we explore some of the fundamentals that underlie the computer methods to find the best fit. The accessibility of microcomputers, starting in the late 1970s, was a great boon for ecological modeling. Many software programs now include optimization routines to automatically find the best fit. New kinds of optimization methods (genetic algorithms, neural networks, simulated annealing) are still being developed. Even so, it is good to understand at least a little bit about how such things are done—on occasion, it might even be easier for you to do it yourself than rely on a built-in routine. But keep in mind that each of us must find the right balance between knowing how to use resources and how to develop them. Here we provide introductory material to give you an understanding of how nonlinear minimization methods work and illustrate some of the simpler methods. We strongly recommend that you purchase *Numerical Recipes: The Art of Scientific Computing* (Press et al. 1986) and *Handbook of Mathematical Functions* (Abramowitz and Stegun 1965). These will stand by you.

## DIRECT SEARCH AND GRAPHICS

Systems with fewer than three parameters are best solved by direct search if the allowable range of the parameters is not too large. Search systematically over possible values of the parameter(s) of interest, and print or plot the relationship between the parameter(s) and the goodness of fit. As a general rule, always plot the goodness of fit if you can. Just as we preached KNOW YOUR DATA, you should also

# Understand the Shape of the Goodness of Fit.

Most data analysis and computation is exploratory; you should expect to conduct 10–100 runs of a computer program, exploring options and debugging for every run that will appear in a table or a figure of a report of the work. Therefore it is important that you see and understand as much as possible about what is happening in the fitting of the model to the data. We recommend that you look at results in real time; see it on the screen as the program runs, rather than run the program, put the output in a file, and then view the output with another program.

For example, the following pseudocode can be used to generate contour plots of the abundance model discussed in Chapter 7.

---

Pseudocode 11.1

1. Read in the observed densities and index of abundance.
2. Determine the lower limit, upper limit, and step size for $q$.
3. Determine the lower limit, upper limit, and step size for $p$.
4. Set $p$ and $q$ at the lower limit; set $r = 0$.
5. Calculate the negative log-likelihood of the data, given estimates of $p$ and $q$.
6. Increment $p$ by its step size and repeat step 5 until the upper limit of $p$ is reached.
7. Increment $q$ by its step size and repeat steps 5 and 6 until the upper limit of $q$ is reached.
8. Output the results in a readable table or contour plot.

---

Here are some hints. First, provide row and column labels of $p$ and $q$ values. These are helpful in interpreting output. Second, input the starting and ending points and the step size of the direct search as parameters, rather than putting specific numbers into the appropriate loops. This makes it much easier to change the region of search. Third, most computer screens offer a resolution of roughly twenty rows and eighty columns. If you allow five columns for each number then you can print out about fifteen columns of numbers in twenty rows. Easy visualization of the shape of a two-dimensional surface thus limits you to fifteen or twenty steps. Fourth, begin with a rough search and then, once you see the general shape of the surface, focus on the region of interest. For instance, using this algorithm with $q$ ranging from 0.2 to 2 in steps of 0.1 and $p$ ranging from $-8$ to 2 in steps of 1 leads to:

| $q$ | Negative Log-Likelihood | | | | | | | | | | |
|------|---|---|---|---|---|---|---|---|---|---|---|
| 0.20 | 999 | 999 | 999 | 999 | 999 | 999 | 999 | 642 | 366 | 284 | 240 |
| 0.30 | 999 | 999 | 999 | 999 | 999 | 999 | 999 | 331 | 251 | 212 | 192 |
| 0.40 | 999 | 999 | 999 | 999 | 999 | 999 | 411 | 227 | 188 | 171 | 167 |
| 0.50 | 999 | 999 | 999 | 999 | 999 | 999 | 215 | 170 | 153 | 151 | 157 |
| 0.60 | 999 | 999 | 999 | 999 | 999 | 305 | 160 | 139 | 136 | 144 | 157 |
| 0.70 | 999 | 999 | 999 | 999 | 999 | 157 | 131 | 125 | 132 | 146 | 165 |
| 0.80 | 999 | 999 | 999 | 999 | 251 | 128 | 118 | 122 | 136 | 156 | 180 |
| 0.90 | 999 | 999 | 999 | 999 | 130 | 115 | 116 | 128 | 148 | 172 | 199 |
| 1.00 | 999 | 999 | 999 | 225 | 116 | **114** | 123 | 140 | 165 | 192 | 222 |
| 1.10 | 999 | 999 | 999 | 121 | 115 | 121 | 136 | 158 | 186 | 216 | 249 |
| 1.20 | 999 | 999 | 217 | 118 | 122 | 135 | 155 | 180 | 211 | 243 | 278 |
| 1.30 | 999 | 999 | 125 | 125 | 135 | 153 | 177 | 206 | 239 | 273 | 310 |
| 1.40 | 999 | 222 | 130 | 138 | 154 | 176 | 203 | 234 | 269 | 306 | 344 |
| 1.50 | 999 | 138 | 142 | 157 | 177 | 202 | 232 | 266 | 302 | 340 | 379 |
| 1.60 | 235 | 149 | 161 | 179 | 203 | 231 | 263 | 299 | 337 | 376 | 417 |
| 1.70 | 159 | 166 | 183 | 205 | 232 | 263 | 297 | 334 | 373 | 414 | 455 |
| 1.80 | 174 | 188 | 209 | 235 | 264 | 296 | 332 | 371 | 411 | 453 | 495 |
| 1.90 | 195 | 214 | 238 | 266 | 297 | 332 | 369 | 409 | 451 | 493 | 536 |
| | $-8$ | $-7$ | $-6$ | $-5$ | $-4$ | $-3$ | $-2$ | $-1$ | 0 | 1 | 2 | $p$ |

We see already that the minimum negative log-likelihood is 114, which is why we highlighted it in bold. The minimum occurs at $q = 1$, $p = -3$. We find the approximate 95% confidence interval by recognizing that a likelihood profile with two free parameters corresponds to a chi-square distribution with two parameters, and the 0.05 level is roughly 6. Thus, the 95% confidence interval includes values of the negative log-likelihood that are less than $114 + 3 = 117$; the 95% confidence region is in the range of $q = 0.90-1.10$ and $p = -4$ to $-2$.

We reset the starting points and step sizes, subtract 100 from the negative log-likelihood in order to show more significant digits, and have

| $q$ | Negative Log-Likelihood | | | | | | | |
|------|------|------|------|------|------|------|------|------|
| 0.80 | 99.0 | 99.0 | 51.0 | 40.8 | 27.8 | 20.7 | 18.1 | 18.6 | 22.2 |
| 0.85 | 99.0 | 99.0 | 44.4 | 27.9 | 19.9 | 16.0 | 16.1 | 18.9 | 24.2 |
| 0.90 | 99.0 | 99.0 | 29.9 | 19.7 | 15.3 | 14.3 | 16.5 | 21.2 | 28.0 |
| 0.95 | 99.0 | 34.2 | 20.9 | 15.2 | **13.4** | 15.0 | 18.8 | 25.4 | 33.5 |
| 1.00 | 99.0 | 23.6 | 15.7 | 13.4 | 13.8 | 17.6 | 23.0 | 31.1 | 40.4 |
| 1.05 | 28.6 | 17.9 | 14.0 | 14.0 | 16.7 | 22.0 | 29.0 | 38.3 | 48.7 |
| 1.10 | 21.1 | 15.4 | 14.7 | 16.7 | 21.3 | 27.9 | 36.4 | 46.7 | 58.2 |
| 1.15 | 18.1 | 15.7 | 17.4 | 21.0 | 27.4 | 35.3 | 45.0 | 56.3 | 68.8 |
| 1.20 | 17.9 | 18.5 | 21.9 | 27.2 | 34.9 | 43.8 | 54.8 | 66.9 | 80.3 |
| 1.25 | 20.0 | 23.0 | 27.9 | 34.8 | 43.7 | 53.5 | 65.6 | 78.5 | 92.7 |
| | $-5.0$ | $-4.5$ | $-4.0$ | $-3.5$ | $-3.0$ | $-2.5$ | $-2.0$ | $-1.5$ | $-1.0$ | $p$ |

The minimum in this table is 13.4, corresponding to a negative log-likelihood of 113.4. Thus, the 95% confidence interval will be roughly bounded by $13.4 + 3.0 = 16.4$. Were this table printed on paper, we could quickly trace the rough 95% confidence bounds (you might want to do so on a photocopy of this page).

This direct search is limited to two parameters. For models with three parameters, one can begin by systematically fixing one parameter and then doing a direct search over the other two. This provides a sense of the shape of the goodness-of-fit surface.

The most important item to check in the goodness-of-fit surface is the presence of multiple minima. Almost all function minimization routines have difficulties, or will fail outright, if there are multiple minima. The algorithm finds a minimum, discovers that the goodness of fit gets worse in all directions, and concludes it has found the best spot. After you have checked for multiple minima, study the general shape of the likelihood surface. Be especially aware of possible problems, such as very long flat valleys or discontinuities. If there are large regions of parameter space where the goodness of fit is very flat, many algorithms will have difficulty, or may fail, and interpretation of the results is very difficult. Similarly, discontinuities in the goodness of fit may confuse many search algorithms. We emphasize the need to check for multiple minima, flat valleys, and discontinuities, because this is where most algorithms fail. However, if your problem is simple, then you can use direct searching by looping over parameter values and printing or plotting the output.

## NEWTON'S METHOD AND GRADIENT SEARCH

When maximizing a likelihood function or minimizing a sum of squares or negative log-likelihood, we often need to find the value of the parameter $p$ (or parameters) that satisfies a nonlinear equation. For example, if $\mathcal{L}\{\text{data}|p\}$ is the likelihood of the data, given a particular value for the parameter, the MLE for the parameter is found by solving

$$\frac{d\mathcal{L}\{\text{data}|p\}}{dp} = 0. \tag{11.1}$$

Since the logarithm is a monotonic function, we can also find the MLE for the parameter by setting the derivative of the logarithm of the likelihood equal to 0. Although many modern software packages contain maximization and minimization routines, we believe that it is worthwhile to understand a little bit of how these work. Consequently, we now

267

describe some simple methods for solving equations such as Equation 11.1 when there are one or two unknown parameters. One of us (M.M.) uses these methods frequently, rather than relying on built-in algorithms (e.g., Mangel and Adler 1994).

For simplicity of notation, we shall consider having to solve

$$H(p) = 0, \tag{11.2}$$

where $H(p)$ is the appropriate equation (e.g., the derivative of the log-likelihood function or the derivative of the sum of squared deviations).

Suppose that $p_t$ is the most likely value of the parameter, which means that

$$H(p_t) = 0. \tag{11.3}$$

Imagine a value $p$ of the parameter other than $p_t$. If we Taylor-expand the function $H(p)$ around $p$ and keep only the first two terms,

$$H(p_t) = H(p) + H'(p)(p_t - p) + \cdots, \tag{11.4}$$

where $H'(p) = dH/dp$ is the first derivative and the "$+\ldots$" denotes all the higher-order terms in the Taylor expansion. The left-hand side of Equation 11.4 is 0, because of the definition of $p_t$, so we solve it for $p_t$:

$$p_t = p - \frac{H(p)}{H'(p)} + \cdots. \tag{11.5}$$

Obviously, this equation cannot be true in general, because we have ignored all the extra terms in the Taylor expansion (the terms contained in "$+ \cdots$"). If these terms are ignored, then Equation 11.5 is correct when $p = p_t$, since it becomes $p_t = p_t$!

For other values of $p$, Equation 11.5 suggests an *iteration equation* for getting to the solution. That is, we replace $p$

with the current guess $p_n$ for the parameter and $p_t$ with the next guess $p_{n+1}$ for the parameter, and write

$$p_{n+1} = p_n - \frac{H(p_n)}{H'(p_n)}. \tag{11.6}$$

Again, note that if $p_n$ is equal to $p_t$, then $p_{n+1}$ will also equal the true value of the parameter. Equation 11.6 is called Newton's method. To get the method going, one picks a first choice $p_1$ and then iterates. Under a wide variety of general conditions (Press et al. 1986), the method converges to $p_t$. The following pseudocode can be used to implement Newton's method for one parameter:

---

Pseudocode 11.2

1. Input a first guess for the parameter $p_1$ and a cutoff level $C$, such that if $|H(p)| < C$, the program stops. Set $n = 1$.
2. Evaluate $H(p_n)$ and $H'(p_n)$.
3. If $|H(p_n)| < C$, go to step 4. Otherwise, set $p_{n+1} = p_n - H(p_n/H'(p_n)$ and return to step 2.
4. Check that you have found the correct kind of extremum (maximum or minimum, as appropriate) by comparing the value of the function to nearby values of the parameter.

---

When there are two unknown parameters, the starting point is the pair of equations

$$H_1(p_{1t},p_{2t}) = 0,$$
$$H_2(p_{1t},p_{2t}) = 0. \tag{11.7}$$

The natural way in which these equations would arise is easily seen if we consider a log-likelihood that depends on two parameters, so that the negative log-likelihood is $\mathbf{L}\{\text{data}|p_1,p_2\}$. The MLE values of the parameters then satisfy

$$\partial\mathbf{L}\{\text{data}|p_1,p_2\}/\partial p_1 = H_1(p_1,p_2) = 0,$$
$$\partial\mathbf{L}\{\text{data}|p_1,p_2\}/\partial p_2 = H_2(p_1,p_2) = 0, \tag{11.8}$$

for which the solutions are assumed to be $p_{1t}$ and $p_{2t}$. Proceeding as before from any value $(p_1,p_2)$, we have

$$H_1(p_{1t},p_{2t}) = 0 = H_1(p_1,p_2)$$
$$+ \frac{\partial H_1}{\partial p_1} (p_{1t} - p_1)$$
$$+ \frac{\partial H_1}{\partial p_2} (p_{2t} - p_2) + \cdots,$$

$$H_2(p_{1t},p_{2t}) = 0 = H_2(p_1,p_2)$$
$$+ \frac{\partial H_2}{\partial p_1} (p_{1t} - p_1)$$
$$+ \frac{\partial H_2}{\partial p_2} (p_{2t} - p_2) + \cdots, \tag{11.9}$$

where the partial derivatives are evaluated at $(p_1,p_2)$.

We rewrite these equations as a pair of linear equations for $p_{1t}$ and $p_{2t}$, analogous to Equation 11.5:

$$\frac{\partial H_1}{\partial p_1} p_{1t} + \frac{\partial H_1}{\partial p_2} p_{2t} = \frac{\partial H_1}{\partial p_1} p_1$$
$$+ \frac{\partial H_1}{\partial p_2} p_2 - H_1(p_1,p_2) + \cdots,$$

$$\frac{\partial H_2}{\partial p_1} p_{1t} + \frac{\partial H_2}{\partial p_2} p_{2t} = \frac{\partial H_2}{\partial p_1} p_1$$
$$+ \frac{\partial H_2}{\partial p_2} p_2 - H_2(p_1,p_2) + \cdots. \tag{11.10}$$

The analogy of the iteration equation, Equation 11.6, is

$$\frac{\partial H_1}{\partial p_1} p_{1,n+1} + \frac{\partial H_1}{\partial p_2} p_{2,n+1} = \frac{\partial H_1}{\partial p_1} p_{1,n}$$
$$+ \frac{\partial H_1}{\partial p_2} p_{2,n} - H_1(p_{1,n},p_{2,n}),$$

$$\frac{\partial H_2}{\partial p_1} p_{1,n+1} + \frac{\partial H_2}{\partial p_2} p_{2,n+1} = \frac{\partial H_2}{\partial p_1} p_{1,n}$$
$$+ \frac{\partial H_2}{\partial p_2} p_{2,n} - H_2(p_{1,n},p_{2,n}). \tag{11.11}$$

Equation 11.11 is a pair of linear equations for $(p_{1,n+1}, p_{2,n+1})$, given the current values of the parameters, the two functions, and the four partial derivatives. These are solved by the standard methods of linear algebra, familiar from experience with the pair of equations $ax + by = c$ and $dx + ey = f$. As an exercise, we encourage you to write a program to solve such a pair of linear equations, for arbitrary parameters $a$, $b$, $c$, $d$, $e$, and $f$, which are treated as inputs.

The method we just described usually works, especially if the likelihood or sum of squares is nicely behaved and not too irregular. But you should know that sometimes these methods do not work. The situation in which they will fail can be seen by looking at Equation 11.6: if there is any chance that the derivative will get close to 0, then there will be problems. (Of course, if the derivative hits 0, there are certainly big problems.) To find out if the method is likely to work, we recommend that before starting the iteration you check the value of the derivative over a reasonable range of parameter values. Similarly, the linear equations in Equation 11.11 may not have a solution. We recommend the same kind of check here. Ways of dealing with these pathological cases are described in texts on numerical methods. Press et al. (1986) is a good starting point.

## NONGRADIENT METHODS: AVOIDING THE DERIVATIVE

There are alternatives to taking derivatives. One of our favorite methods for doing this, when the likelihood function has a single extremum, is the golden section search (Wismer and Chattergy 1978) to find the value of the parameter that makes a function $H(p)$ an extremum. Although this method only works for problems with one parameter, it is worth knowing. We demonstrate the method for finding a maximum.

271

Assume that the value of the parameter is in the interval $p_L \leq p \leq p_U$, where $p_L$ and $p_U$ are specified by the user. The variables used in the method are:

$p_{L,n}$ = lower limit of the range of parameter values on the $n^{th}$ iteration

$p_{U,n}$ = upper limit of the range of parameter values on the $n^{th}$ iteration

$p_{1,n}$ and $p_{2,n}$ = two "test values" (see below) of the parameter on the $n^{th}$ iteration.

The two test values are chosen according to

$$p_{2,n} = p_{U,n} - (1 - \lambda)(p_{U,n} - p_{L,n}),$$

$$p_{1,n} = p_{L,n} + (1 - \lambda)(p_{U,n} - p_{L,n}), \qquad (11.12)$$

where $\lambda = (\sqrt{5} - 1)/2$ is a solution of the equation $\lambda^2 + \lambda - 1 = 0$; for the motivation of this choice see Wismer and Chattergy (1978, 127–28). We then evaluate the function at each of these test points. If $H(p_{2,n}) < H(p_{1,n})$ then

- Set $p_{U,n+1} = p_{2,n}$
- Set $p_{L,n+1} = p_{L,n}$
- Set $p_{2,n+1} = p_{1,n}$
- Determine $p_{1,n+1}$ from Equation 11.12

If $H(p_{2,n}) > H(p_{1,n})$ then

- Set $p_{U,n+1} = p_{U,n}$
- Set $p_{L,n+1} = p_{1,n}$
- Set $p_{1,n+1} = p_{2,n}$
- Determine $p_{2,n+1}$ from Equation 11.12

This description actually should suffice as a pseudocode, but as an exercise we encourage you to write out the pseudocode itself. If we apply this method to maximize the $H(p) = -(p - 1.235)^2 + 0.78p + 0.2$, with $-5 < p < 5$, the output is:

| $n$ | $p_{1n}$ | $H(p_{1n})$ | $p_{2n}$ | $H(p_{2n})$ |
|---|---|---|---|---|
| 1 | $-1.180\ 34$ | $-6.554\ 53$ | 1.180 34 | 1.117 68 |
| 2 | 1.180 34 | 1.117 68 | 2.639 32 | 0.286 554 |
| 3 | 0.278 64 | $-0.497\ 284$ | 1.180 34 | 1.117 68 |
| 4 | 1.180 34 | 1.117 68 | 1.737 62 | 1.302 72 |
| 5 | 1.737 62 | 1.302 72 | 2.082 04 | 1.106 52 |
| 6 | 1.524 76 | 1.305 35 | 1.737 62 | 1.302 72 |
| 7 | 1.393 2 | 1.261 67 | 1.524 76 | 1.305 35 |
| 8 | 1.524 76 | 1.305 35 | 1.606 06 | 1.315 04 |
| 9 | 1.606 06 | 1.315 04 | 1.656 31 | 1.314 42 |
| 10 | 1.575 01 | 1.312 9 | 1.606 06 | 1.315 04 |
| 11 | 1.606 06 | 1.315 04 | 1.625 26 | 1.315 4 |
| 12 | 1.625 26 | 1.315 4 | 1.637 12 | 1.315 25 |
| 13 | 1.617 93 | 1.315 35 | 1.625 26 | 1.315 4 |
| 14 | 1.625 26 | 1.315 4 | 1.629 79 | 1.315 38 |
| 15 | 1.622 46 | 1.315 39 | 1.625 26 | 1.315 4 |
| 16 | 1.625 26 | 1.315 4 | 1.626 99 | 1.315 4 |
| 17 | 1.624 19 | 1.315 4 | 1.625 26 | 1.315 4 |
| 18 | 1.625 26 | 1.315 4 | 1.625 92 | 1.315 4 |
| 19 | 1.624 85 | 1.315 4 | 1.625 26 | 1.315 4 |
| 20 | 1.624 6 | 1.315 4 | 1.624 85 | 1.315 4 |

and we see that after twenty iterations, the method has pretty much converged.

## THE ART OF FITTING

Obtaining the best goodness of fit is as much art as science and is a skill that must be learned. In particular, the master of this business conducts searches efficiently and understands why something fails. Allow us an anecdote. We once knew a young ecologist who attempted to estimate the movement rates between different areas based on mark recapture of thousands of individuals. All the ingredients were there: an important problem, a good hypothesis, and lots of data. After a year of trying to get his computer program to work, he gave up and dropped the project, because he could not get the program to produce sensible answers. We believe that this work was never completed, and he effectively wasted a year of his life. He made one simple mistake—he attempted to fit the model to his data.

What is wrong, you may ask, with fitting the model to the data—isn't that what the ecological detective is supposed to do? Yes, but first the detective has to fit the model to data where the correct answer is known. All but the simplest nonlinear minimization problems are complex: the function must be coded correctly, the data must be input correctly, the likelihood must be programmed correctly, and the search algorithm has to be given good starting estimates. It is impossible to be certain that a computer program is working unless it produces a known result from a specific set of data, and, as we mentioned before, even then it may not work with real data. The ecological detective does not know the correct answer before starting and therefore must debug the program very carefully. Complex programs should be debugged in a systematic, step by step process. The three key steps are:

1. Generate deterministic data from the model and check the program with these data. Given deterministic data from the exact model, if the program does not converge to the correct parameters, there is clearly an error in the program. Some methods will not work when all observations exactly match the predictions; thus sometimes you have to add very small random variables to the data even in the deterministic step.

2. Add observation uncertainty to the simulated data, and observe how well the estimation procedure works with Monte Carlo data. In general, if the program worked before, it is likely that you have programmed it correctly. This step of Monte Carlo testing is also important to determine if there are biases in the estimation procedure.

274

3. Fit the real data—but remember, if you skip the first two steps, you have no way of knowing that the answer you obtain is actually correct!

Once you have acquired some experience at this business, it will be tempting to go right to step 3. Don't do it! Steps 1 and 2 are quick and can save you lots of time—as well as embarrassment if you get wrong results and do not notice this until your presentation.

We illustrate this three-step method with a problem commonly encountered in the study of the abundance of animals. If the numbers of animals grows logistically, and we cannot observe the actual number but have an abundance index that is proportional to population size, the model is

$$N_{t+1} = N_t + rN_t \left( 1 - \frac{N_t}{K} \right) - C_t,$$

$$I_t = qN_t. \tag{11.13}$$

Here $N_t$ is the number of individuals at time $t$, $I_t$ is the index of abundance at time $t$, $C_t$ is the catch at time $t$, $r$ is the maximum per capita rate of growth, $K$ is the carrying capacity, and $q$ is the constant of proportionality between the abundance and the index of abundance.

Assuming that the stock is initially at carrying capacity, we generate ten data points (catch and index) using the following pseudocode:

---

Pseudocode 11.3

1. Input the true parameters $K_{true}$, $r_{true}$, and $q_{true}$.
2. Let $N_0 = K$.
3. Specify the catch over each of the ten periods; i.e., give $C_t$ for $t = 0$ to $t = 9$.
4. Loop over $t$, from $t = 1$ to $t = 10$, and determine $N_t$ and $I_t$ from Equation 11.13, with $K = K_{true}$, $r = r_{true}$, and $q = q_{true}$.

---

To estimate the parameters in the model from the data, we must introduce a goodness of fit. If the measure of goodness of fit is the sum of squared deviations between the logarithm of the predicted index of abundance and the logarithm of the observed index of abundance, we continue the pseudocode:

---

Pseudocode 11.3 (continued)

5. Once again, set $K = K_{true}$, $r = r_{true}$, and $q = q_{true}$. Specify $N_0 = K$ and generate a predicted index of abundance $I_{pre,t}$ using Equation 11.13.

6. Determine the sum of squared deviations

$$\mathcal{S} = \sum_{t=1}^{10} [\log(I_{pre,t}) - \log(I_t)]^2.$$

---

If we have done everything correctly, the value of the sum of squared deviations should be 0, since we are using the true parameters.

Next, we modify the pseudocode to search over a range of parameter values, for the case where we do not know the true parameters:

---

Pseudocode 11.3 (continued)

7. Input the range of allowable values for $K$, $r$, and $q$, and the increments in which this range will be searched. Set $\mathcal{S}_{min}$ to a large value.

8. Cycle over $K$, $r$, and $q$ in the appropriate increments and repeat steps 4 and 5 for each value of $K$, $r$, and $q$. If the value of the sum of squared deviations from step 5 is less than the current value of $\mathcal{S}_{min}$, set $\mathcal{S}_{min}$ to this value of $\mathcal{S}$ and set $K^*$, $r^*$, and $q^*$ to the current parameters.

9. Use a graphics routine to plot the goodness-of-fit

measure as a function of the parameters. That is, think
of the sum of squared deviations as a function
$\mathcal{S}_{\min}(K,r,q)$ of the parameter values and plot
$\mathcal{S}_{\min}(K,r^*,q^*)$ versus $K$, $\mathcal{S}_{\min}(K^*,r,q^*)$ versus $r$, and
$\mathcal{S}_{\min}(K^*,r^*,q)$ versus $q$.

---

This pseudocode completes step 1. Note that it might not
lead to the true parameters. For example, the true parameters might be out of the range that you allow for the parameters, or might not be reachable because of the limits and
the increments you have chosen. We leave it to you to modify the pseudocode for steps 2 and 3.

Only now are we ready to try fitting the model to real
data. If you follow these steps carefully, and make sure that
each step is working, it should take only a few hours to get a
minimization program running. Modern interactive computer programming languages greatly facilitate the work of
the ecological detective. If you are using computer software
that does not allow real-time graphics and interactive debugging, you are working too hard. Take a day or two to learn
how to use a software systems such as QuickBASIC™, True-
BASIC™, TurboBASIC™, or TurboPASCAL™. More sophisti-
cated systems, $C^{++}$ (for computation) and $S^+$ (for graphics),
take more time to learn, but may be worthwhile investments.

There are many packaged programs now available that
will "do the programming for you." Spreadsheets such as
Excel™ have built-in nonlinear minimization packages that
are adequate for many problems. Packages such as Mathe-
matica™, Mathcad™, and Systat™ also do function minimization and many people swear by them. We do almost all of
our work in BASIC (although R.H. works largely in Excel).
However, so long as the software allows real-time graphics,
nonlinear function minimization, and interactive debugging, you can't go too far wrong.

As you start using more complex models, you will often

277

want to examine nested models. For example, imagine that we have a code working for a particular problem. Now, if we wanted to add another parameter, or perhaps treat some of the parameters as fixed, we have to change the code. This is not too much trouble when dealing with two or three parameters, but when dealing with a dozen parameters, it is essential to write the program so that you can turn a particular variable on or off, and your program will automatically initialize the list of parameters appropriately.

## HINTS FOR SPECIAL PROBLEMS

### *Constrained Parameters*

Most minimization algorithms assume that the parameters are unconstrained, and will search over real values of each parameter to minimize the function. However, it often occurs that the parameter values are only meaningful over certain ranges. For example, we may require that a parameter is positive, or that it must be between 0 and 1. There are several ways to work with these types of problems. First, it may be possible to scale the parameters so that the algorithm searches over all possible real numbers, but the transformation of the parameter, the one used in the goodness of fit, varies over the appropriate range. For example, suppose that the real parameter is constrained to $0 < p < 1$ in the model, but that the search routine requires $-\infty < p_s < \infty$. One way of dealing with this is to search over all values of $p_s$, but in the model set

$$p = [(\pi/2) + \arctan(p_s)]/\pi. \qquad (11.14)$$

To understand this choice, recall that the function $y = \arctan(x)$ takes $y$ values in $[0, \pi/2]$ if $x > 0$ and $y$ values in $[-\pi/2, 0)$ if $x$ is negative. Thus, if we then add $\pi/2$ and normalize by $\pi$, we have a transformation which takes the value of $p_s$ into the constrained interval $[0,1]$. This method can be adapted to any constrained parameter range, and it provides great benefits in computing time over methods that have the ability to do constraints directly.

Another approach, and often the simplest, is to penalize the goodness of fit if the search algorithm tries to use a parameter value outside the desired range. If the value of the parameter is outside the desired range, set the goodness of fit to a very large number. This is a brute force method that will work in many cases, but can cause very poor behavior under some circumstances. It is generally better to have the penalty be a smooth function of the parameter. Discontinuous penalties may cause the search algorithm to "bounce around" the threshold of the penalty.

### Checks on Convergence

It is good practice always to check the program to make sure that the convergence to the best parameters is repeatable. To do so, let the program converge to the parameters that minimize the function. Then restart it from a totally different initial guess and see if it finds the same minimum and the same parameters. If the program converges to a minimum that is lower than the initial minimum, you have a goodness-of-fit surface that has multiple peaks (or a programming error), and you must carefully explore the parameter space.

### Confounded Parameters

It you are not careful, the model may be written so that two parameters are confounded. As we saw previously, a simple example is the binomial distribution with known $N$ and unknown $p$, and data consisting of $k$ of $N$ successes. Then a good estimate of $p$ is $k/N$. If, on the other hand, $p$ is known but $N$ is not, a good estimate of $N$ is the integer part of $k/p$. However, if both $p$ and $N$ are unknown, great difficulties are encountered. The result is that the search algorithms get confused in what is, in effect, a very long valley, and fail to converge.

# CHAPTER ELEVEN

*Truly Large Problems*

For truly large problems (more than twenty parameters), computer implementation becomes critical. The algorithms we have used in this book are not designed for such large problems, although they have been used on problems of 30–40 parameters. In cases such as this, it is time to consult the experts in numerical methods.

# "The Method of Multiple Working Hypotheses" by T. C. Chamberlain

There are two fundamental modes of study. The one is an attempt to follow by close imitation the processes of previous thinkers and to acquire the results of their investigations by memorizing. It is study of a merely secondary, imitative, or acquisitive nature. In the other mode the effort is to think independently, or at least individually. It is primary or creative study. The endeavor is to discover new truth or to make a new combination of truth or at least to develop by one's own effort an individualized assemblage of truth. The endeavor is to think for one's self, whether the thinking lies wholly in the fields of previous thought or not. It is not necessary to this mode of study that the subject-matter should be new. Old material may be reworked. But it is essential that the process of thought and its results be individual and independent, not the mere following of previous lines of thought ending in predetermined results. The demonstration of a problem in Euclid precisely as laid down is an illustration of the former; the demonstration of the same proposition by a method of one's own or in a manner distinctively individual is an illustration of the latter, both lying entirely within the realm of the known and old.

Thomas C. Chamberlain was a geologist, president of the University of Wisconsin, president of the American Association for the Advancement of Science, Director of the Walker Museum at the University of Chicago and founder of the *Journal of Geology*. The paper that we reprint here was first published in 1890 in *Science* 15:92 and then later in the *Journal of Geology* 5:837–48 (1897).

Creative study however finds its largest application in those subjects in which, while much is known, more remains to be learned. The geological field is preeminently full of such subjects, indeed it presents few of any other class. There is probably no field of thought which is not sufficiently rich in such subjects to give full play to investigative modes of study.

Three phases of mental procedure have been prominent in the history of intellectual evolution thus far. What additional phases may be in store for us in the evolutions of the future it may not be prudent to attempt to forecast. These three phases may be styled the method of the ruling theory, the method of the working hypothesis, and the method of multiple working hypotheses.

In the earlier days of intellectual development the sphere of knowledge was limited and could be brought much more nearly than now within the compass of a single individual. As a natural result those who then assumed to be wise men, or aspired to be thought so, felt the need of knowing, or at least seeming to know, all that was known, as a justification of their claims. So also as a natural counterpart there grew up an expectancy on the part of the multitude that the wise and the learned would explain whatever new thing presented itself. Thus pride and ambition on the one side and expectancy on the other joined hands in developing the putative all-wise man whose knowledge boxed the compass and whose acumen found an explanation for every new puzzle which presented itself. Although the pretended compassing of the entire horizon of knowledge has long since become an abandoned affectation, it has left its representatives in certain intellectual predilections. As in the earlier days, so still, it is a too frequent habit to hastily conjure up an explanation for every new phenomenon that presents itself. Interpretation leaves its proper place at the end of the intellectual procession and rushes to the forefront. Too often a theory is promptly born and evidence hunted up to fit in afterward. Laudable as the effort at explanation is in its proper place, it is an almost certain source of confusion and error when it runs before a serious inquiry into the phenomenon itself. A strenuous endeavor to find out precisely what the phenomenon really is should take the lead and crowd back the question, commendable at a later stage, "How came this so?" First the full facts, then the interpretation thereof, is the normal order.

The habit of precipitate explanation leads rapidly on to the birth of general theories.[1] When once an explanation or special theory has been offered for a given phenomenon, self-consistency prompts to the offering of the same explanation or theory for like phenomena when they present themselves and there is soon developed a general theory explanatory of a large class of phenomena similar to the original one. In support of the general theory there may not be any further evidence or investigation than was involved in the first hasty conclusion. But the repetition of its application to new phenomena, though of the same kind, leads the mind insidiously into the delusion that the theory has been strengthened by additional facts. A thousand applications of the supposed principle of levity to the explanation of ascending bodies brought no increase of evidence that it was the true theory of the phenomena, but it doubtless created the impression in the minds of ancient physical philosophers that it did, for so many additional facts seemed to harmonize with it.

For a time these hastily born theories are likely to be held in a tentative way with some measure of candor or at least some self-illusion of candor. With this tentative spirit and

[1] I use the term theory here instead of hypothesis because the latter is associated with a better controlled and more circumspect habit of the mind. This restrained habit leads to the use of the less assertive term hypothesis, while the mind in the habit here sketched more often believes itself to have reached the higher ground of a theory and more often employs the term theory. Historically also I believe the word theory was the term commonly used at the time this method was predominant.

measurable candor, the mind satisfies its moral sense and deceives itself with the thought that it is proceeding cautiously and impartially toward the goal of ultimate truth. It fails to recognize that no amount of provisional holding of a theory, no amount of application of the theory, so long as the study lacks in incisiveness and exhaustiveness, justifies an ultimate conviction. It is not the slowness with which conclusions are arrived at that should give satisfaction to the moral sense, but the precision, the completeness and the impartiality of the investigation.

It is in this tentative stage that the affections enter with their blinding influence. Love was long since discerned to be blind and what is true in the personal realm is measurably true in the intellectual realm. Important as the intellectual affections are as stimuli and as rewards, they are nevertheless dangerous factors in research. All too often they put under strain the integrity of the intellectual processes. The moment one has offered an original explanation for a phenomenon which seems satisfactory, that moment affection for his intellectual child springs into existence, and as the explanation grows into a definite theory his parental affections cluster about his offspring and it grows more and more dear to him. While he persuades himself that he holds it still as tentative, it is none the less lovingly tentative and not impartially and indifferently tentative. So soon as this parental affection takes possession of the mind, there is apt to be a rapid passage to the unreserved adoption of the theory. There is then imminent danger of an unconscious selection and of a magnifying of phenomena that fall into harmony with the theory and support it and an unconscious neglect of phenomena that fail of coincidence. The mind lingers with pleasure upon the facts that fall happily into the embrace of the theory, and feels a natural coldness toward those that assume a refractory attitude. Instinctively there is a special searching-out of phenomena that support it, for the mind is led by its desires. There springs up also unwittingly a pressing of the theory to make it fit the facts and a pressing of the facts to make them fit the theory. When these biasing tendencies set in, the mind rapidly degenerates into the partiality of paternalism. The search for facts, the observation of phenomena and their interpretation are all dominated by affection for the favored theory until it appears to its author or its advocate to have been overwhelmingly established. The theory then rapidly rises to a position of control in the processes of the mind and observation, induction and interpretation are guided by it. From an unduly favored child it readily grows to be a master and leads its author whithersoever it will. The subsequent history of that mind in respect to that theme is but the progressive dominance of a ruling idea. Briefly summed up, the evolution is this: a premature explanation passes first into a tentative theory, then into an adopted theory, and lastly into a ruling theory.

When this last stage has been reached, unless the theory happens perchance to be the true one, all hope of the best results is gone. To be sure truth may be brought forth by an investigator dominated by a false ruling idea. His very errors may indeed stimulate investigation on the part of others. But the condition is scarcely the less unfortunate.

As previously implied, the method of the ruling theory occupied a chief place during the infancy of investigation. It is an expression of a more or less infantile condition of the mind. I believe it is an accepted generalization that in the earlier stages of development the feelings and impulses are relatively stronger than in later stages.

Unfortunately the method did not wholly pass away with the infancy of investigation. It has lingered on, and reappears in not a few individual instances at the present time. It finds illustration in quarters where its dominance is quite unsuspected by those most concerned.

The defects of the method are obvious and its errors grave. If one were to name the central psychological fault, it might be stated as the admission of intellectual affection to the place that should be dominated by impartial, intellectual rectitude alone.

So long as intellectual interest dealt chiefly with the intangible, so long it was possible for this habit of thought to survive and to maintain its dominance, because the phenomena themselves, being largely subjective, were plastic in the hands of the ruling idea; but so soon as investigation turned itself earnestly to an inquiry into natural phenomena whose manifestations are tangible, whose properties are inflexible, and whose laws are rigorous, the defects of the method became manifest and an effort at reformation ensued. The first great endeavor was repressive. The advocates of reform insisted that theorizing should be restrained and the simple determination of facts should take its place. The effort was to make scientific study statistical instead of causal. Because theorizing in narrow lines had led to manifest evils theorizing was to be condemned. The reformation urged was not the proper control and utilization of theoretical effort but its suppression. We do not need to go backward more than a very few decades to find ourselves in the midst of this attempted reformation. Its weakness lay in its narrowness and its restrictiveness. There is no nobler aspiration of the human intellect than the desire to compass the causes of things. The disposition to find explanations and to develop theories is laudable in itself. It is only its ill-placed use and its abuse that are reprehensible. The vitality of study quickly disappears when the object sought is a mere collocation of unmeaning facts.

The inefficiency of this simply repressive reformation becoming apparent, improvement was sought in the method of the working hypothesis. This has been affirmed to be *the* scientific method. But it is rash to assume that any method is *the* method, at least that it is the ultimate method. The working hypothesis differs from the ruling theory in that it is used as a means of determining facts rather than as a proposition to be established. It has for its chief function the suggestion and guidance of lines of inquiry; the inquiry being made, not for the sake of the hypothesis, but for the sake of the facts and their elucidation. The hypothesis is a mode rather than an end. Under the ruling theory, the stimulus is directed to the finding of facts for the support of the theory. Under the working hypothesis, the facts are sought for the purpose of ultimate induction and demonstration, the hypothesis being but a means for the more ready development of facts and their relations.

It will be observed that the distinction is not such as to prevent a working hypothesis from gliding with the utmost ease into a ruling theory. Affection may as easily cling about a beloved intellectual child when named an hypothesis as if named a theory, and its establishment in the one guise may become a ruling passion very much as in the other. The historical antecedents and the moral atmosphere associated with the working hypothesis lend some good influence however toward the preservation of its integrity.

Conscientiously followed, the method of the working hypothesis is an incalculable advance upon the method of the ruling theory; but it has some serious defects. One of these takes concrete form, as just noted, in the ease with which the hypothesis becomes a controlling idea. To avoid this grave danger, the method of multiple working hypotheses is urged. It differs from the simple working hypothesis in that it distributes the effort and divides the affections. It is thus in some measure protected against the radical defect of the two other methods. In developing the multiple hypotheses, the effort is to bring up into review every rational explanation of the phenomenon in hand and to develop every tenable hypothesis relative to its nature, cause or origin, and to give to all of these as impartially as possible a working form and a due place in the investigation. The investigator thus becomes the parent of a family of hypotheses; and by his parental relations to all is morally forbidden to fasten his affections unduly upon any one. In the very nature of the case, the chief danger that springs from affection is counteracted. Where some of the hypotheses have been already proposed and used, while others are the investigator's own creation. A natural difficulty arises, but the right use of the method requires the impartial adoption of all alike into the working family. The investigator thus at the outset puts himself in cordial sympathy and in parental relations (of adoption, if not of authorship) with every hypothesis that is at all applicable to the case under investigation. Having thus neutralized so far as may be the partialities of his emotional nature, he proceeds with a certain natural and enforced erectness of mental attitude to the inquiry, knowing well that some of his intellectual children (by birth or adoption) must needs perish before maturity, but yet with the hope that several of them may survive the ordeal of crucial research, since it often proves in the end that several agencies were conjoined in the production of the phenomena. Honors must often be divided between hypotheses. One of the superiorities of multiple hypotheses as a working mode lies just here. In following a single hypothesis the mind is biased by the presumptions of its method toward a single explanatory conception. But an adequate explanation often involves the coordination of several causes. This is especially true when the research deals with a class of complicated phenomena naturally associated, but not necessarily of the same origin and nature, as for example the Basement Complex or the Pleistocene drift. Several agencies may participate not only but their proportions and importance may vary from instance to instance in the same field. The true explanation is therefore necessarily complex, and the elements of the complex are constantly varying. Such distributive explanations of phenomena are especially contemplated and encouraged by the method of multiple hypotheses and constitute one of its chief merits. For many reasons we are prone to refer phenomena to a single cause.

If naturally follows that when we find an effective agency present, we are predisposed to be satisfied therewith. We are thus easily led to stop short of full results, sometimes short of the chief factors.. The factor we find may not even be the dominant one, much less the full complement of agencies engaged in the accomplishment of the total phenomena under inquiry. The mooted question of the origin of the Great Lake basins may serve as an illustration. Several hypotheses have been urged by as many different students of the problem as the cause of these great excavations. All of these have been pressed with great force and with an admirable array of facts. Up to a certain point we are compelled to go with each advocate. It is practically demonstrable that these basins were river valleys antecedent to the glacial incursion. It is equally demonstrable that there was a blocking up of outlets. We must conclude then that the present basins owe their origin in part to the preexistence of river valleys and to the blocking up of their outlets by drift. That there is a temptation to rest here, the history of the question shows. But on the other hand it is demonstrable that these basins were occupied by great lobes of ice and were important channels of glacial movement. The leeward drift shows much material derived from their bottoms. We cannot therefore refuse assent to the doctrine that the basins owe something to glacial excavation. Still again it has been urged that the earth's crust beneath these basins was flexed downward by the weight of the ice load and contracted by its low temperature and that the basins owe something to crustal deformation. This third cause tallies with certain features not readily explained by the others. And still it is doubtful whether all these combined constitute an adequate explanation of the phenomena. Certain it is, at least, that the measure of participation of each must be determined before a satisfactory elucidation can be reached. The full solution therefore involves not only the recognition of multiple participation but an estimate of the measure and mode

289

of each participation. For this the simultaneous use of a full staff of working hypotheses is demanded. The method of the single working hypothesis or the predominant working hypothesis is incompetent.

In practice it is not always possible to give all hypotheses like places nor does the method contemplate precisely equable treatment. In forming specific plans for field, office or laboratory work it may often be necessary to follow the lines of inquiry suggested by some one hypothesis, rather than those of another. The favored hypothesis may derive some advantage therefrom or go to an earlier death as the case may be, but this is rather a matter of executive detail than of principle.

A special merit of the use of a full staff of hypotheses coordinately is that in the very nature of the case it invites thoroughness. The value of a working hypothesis lies largely in the significance it gives to phenomena which might otherwise be meaningless and in the new lines of inquiry which spring from the suggestions called forth by the significance thus disclosed. Facts that are trivial in themselves are brought forth into importance by the revelation of their bearings upon the hypothesis and the elucidation sought through the hypothesis. The phenomenal influence which the Darwinian hypothesis has exerted upon the investigations of the past two decades is a monumental illustration. But while a single working hypothesis may lead investigation very effectively along a given line, it may in that very fact invite the neglect of other lines equally important. Very many biologists would doubtless be disposed today to cite the hypothesis of natural selection, extraordinary as its influence for good has been, as an illustration of this. While inquiry is thus promoted in certain quarters, the lack of balance and completeness gives unsymmetrical and imperfect results. But if on the contrary all rational hypotheses bearing on a subject are worked coordinately, thoroughness, equipoise, and symmetry are the presumptive results in the very nature of the case.

In the use of the multiple method, the reaction of one hypothesis upon another tends to amplify the recognized scope of each. Every hypothesis is quite sure to call forth into clear recognition new or neglected aspects of the phenomena in its own interests, but ofttimes these are found to be important contributions to the full deployment of other hypotheses. The eloquent expositions of "prophetic" characters at the hands of Agassiz were profoundly suggestive and helpful in the explication of "undifferentiated" types in the hand of the evolutionary theory.

So also the mutual conflicts of hypotheses whet the discriminative edge of each. The keenness of the analytic process advocates the closeness of differentiating criteria, and the sharpness of discrimination is promoted by the coordinate working of several competitive hypotheses.

Fertility in processes is also a natural sequence. Each hypothesis suggests its own criteria, its own means of proof, its own method of developing the truth; and if a group of hypotheses encompass the subject on all sides, the total outcome of means and of methods is full and rich.

The loyal pursuit of the method for a period of years leads to certain distinctive habits of mind which deserve more than the passing notice which alone can be given them here. As a factor in education the disciplinary value of the method is one of prime importance. When faithfully followed for a sufficient time, it develops a mode of thought of its own kind which may be designated the habit of parallel thought, or of complex thought. It is contra-distinguished from the linear order of thought which is necessarily cultivated in language and mathematics because their modes are linear and successive. The procedure is complex and largely simultaneously complex. The mind appears to become possessed of the power of simultaneous vision from different

291

points of view. The power of viewing phenomena analytically and synthetically at the same time appears to be gained. It is not altogether unlike the intellectual procedure in the study of a landscape. From every quarter of the broad area of the landscape there come into the mind myriads of lines of potential intelligence which are received and coordinated simultaneously producing a complex impression which is recorded and studied directly in its complexity. If the landscape is to be delineated in language it must be taken part by part in linear succession.

Over against the great value of this power of thinking in complexes there is an unavoidable disadvantage. No good thing is without its drawbacks. It is obvious upon studious consideration that a complex or parallel method of thought cannot be rendered into verbal expression directly and immediately as it takes place. We cannot put into words more than a single line of thought at the same time, and even in that the order of expression must be conformed to the idiosyncrasies of the language. Moreover the rate must be incalculably slower than the mental process. When the habit of complex or parallel thought is not highly developed there is usually a leading line of thought to which the others are subordinate. Following this leading line the difficulty of expression does not rise to serious proportions. But when the method of simultaneous mental action along different lines is so highly developed that the thoughts running in different channels are nearly equivalent, there is an obvious embarrassment in making a selection for verbal expression and there arises a disinclination to make the attempt. Furthermore the impossibility of expressing the mental operation in words leads to their disuse in the silent processes of thought and hence words and thoughts lose that close association which they are accustomed to maintain with those whose silent as well as spoken thoughts predominantly run in linear verbal courses. There is therefore a certain predisposition on the part of the practitioner of this method to taciturnity. The remedy obviously lies in coordinate literary work.

An infelicity also seems to attend the use of the method with young students. It is far easier, and apparently in general more interesting, for those of limited training and maturity to accept a simple interpretation or a single theory and to give it wide application, than to recognize several concurrent factors and to evaluate these as the true elucidation often requires. Recalling again for illustration the problem of the Great Lake basins, it is more to the immature taste to be taught that these were scooped out by the mighty power of the great glaciers than to be urged to conceive of three or more great agencies working successively in part and simultaneously in part and to endeavor to estimate the fraction of the total results which was accomplished by each of these agencies. The complex and the quantitative do not fascinate the young student as they do the veteran investigator.

The studies of the geologist are peculiarly complex. It is rare that his problem is a simple unitary phenomenon explicable by a single simple cause. Even when it happens to be so in a given instance, or at a given stage of work, the subject is quite sure, if pursued broadly, to grade into some complication or undergo some transition. He must therefore ever be on the alert for mutations and for the insidious entrance of new factors. If therefore there are any advantages in any field in being armed with a full panoply of working hypotheses and in habitually employing them, it is doubtless the field of the geologist.

# References

Abramowitz, M., and I. Stegun (editors). 1965. *Handbook of Mathematical Functions.* Dover, New York.

Adkison, M. 1992. Parameter estimation for models of chaotic time series. Journal of Mathematical Biology. 30:839–52.

Akaike, H. 1973. Information theory and an extension of the maximum likelihood principle. In *2nd International Symposium on Information Theory*, B. N. Petrov and F. Csaki (editors), pp. 268–81. Publishing House of the Hungarian Academy of Sciences, Budapest. Reprinted in 1992 in *Breakthroughs in Statistics*, S. Kotz and N. Johnson (editors), 1:610–24. Springer Verlag, New York.

Akaike, H. 1985. Prediction and entropy. In *A Celebration of Statistics*, A. C. Atkinson and S. E. Fienberg (editors), pp. 1–24. Springer Verlag, New York.

Akaike, H. 1992. Information theory and an extension of the maximum likelihood principle. In *Breakthroughs in Statistics*, S. Kotz and N. Johnson (editors), 1:610–24. Springer Verlag, New York.

Anderson, D. R., K. P. Burnahm, and G. C. White. 1994. AIC model selection in overdispersed capture-recapture data. Ecology 75:1780–93.

Apostolakis, G. 1990. The concept of probability in safety assessments of technological systems. Science 250:1359–64.

Bar-Hillel, M., and R. Falk. 1982. Some teasers concerning conditional probabilities. Cognition 11:109–22.

Bartle, J. A. 1991. Incidental capture of seabirds in the New Zealand subantarctic squid trawl fishery, 1990. Bird Conservation International 351–59.

Berger, J. 1980. *Bayesian Statistics.* Springer Verlag, New York.

Berger, J. O., and D. A. Berry. 1988. Statistical analysis and the illusion of objectivity. American Scientist 76:159–65.

Bernays, E. A., and P. Wege. 1987. Significance levels of inferential statistics and their interpretation: a lesson from feeding deterrent experiments. Annals of the Entomological Society of America 80:9–11.

Beverton, R. J. H., and S. Holt. 1993. *The Dynamics of Exploited Fish Populations.* Chapman and Hall, London and New York.

Bolt, B. A. 1991. Balance of risks and benefits in preparation for earthquakes. Science 251:169–74.

Caddy, J. F., and J. A. Gulland. 1983. Historical patterns of fish stocks. Marine Policy 7:267–78.

Campbell, K., and H. Hofer. 1995. Humans and wildlife: spatial dynamics and zones of interaction. In Sinclair and Arcese (1995).

Carey, J. R. 1991. Establishment of the Mediterranean fruit fly in California. Science 253:1369–73.

Carpenter, S. R. 1990. Large-scale perturbations: opportunities for innovation. Ecology 71:2038–43.

Caswell, H. 1988. Theory and models in ecology: a different perspective. Ecological Modelling 43:33–44.

Caswell, H. 1989. *Matrix Population Models.* Sinauer Associates, Sutherland, Mass.

Chamberlain, T. C. 1897. The method of multiple working hypotheses. Journal of Geology 5:837–48. Reprinted in Science **148**:754–59 (1965).

Charnov, E., and S. Skinner. 1984. Evolution of host selection and clutch size in parasitoid wasps. Florida Entomologist 67:5–21.

Chen, Y., D. A. Jackson, and H. H. Harvey. 1992. A comparison of von Bertalanffy and polynomial functions in modelling fish growth data. Canadian Journal of Fisheries and Aquatic Sciences 49:1228–35.

Clark, C. W. 1985. *Bioeconomic Modelling and Fisheries Management.* Wiley-Interscience, New York.

Clark, C. W. 1990. *Mathematical Bioeconomics.* 2nd ed. Wiley Interscience, New York.

Cohen, J. 1994. The earth is round ($p < .05$). American Psychologist 49:997–1003.

Cohen, J. E. 1995. *How Many People Can the Earth Support?* W. W. Norton and Company, New York.

Collett, D. 1991. *Modelling Binary Data.* Chapman and Hall, New York.

Conway, G. R., N. R. Glass, and J. C. Wilcox. 1970. Fitting nonlinear models to biological data by Marquardt's algorithm. Ecology 51:503–7.

Cordue, P. L., and R. I. C. C. Francis. 1994. Accuracy and choice in risk estimation for fisheries assessment. Canadian Journal of Fisheries and Aquatic Sciences 51:817–29.

Cowell, R. K. 1984. What's new? Community ecology discovers biology. In *A New Ecology. Novel Approaches to Interactive Systems*, P. W. Price, C. N. Slobodchikoff, and W. S. Gaud (editors), pp. 387–96. John Wiley and Sons, New York.

Crick, F. 1988. *What Mad Pursuit.* Basic Books, New York.

Cronin, J. T., and D. R. Strong. 1993. Substantially submaximal oviposition rates by a mymarid egg parasitoid in the laboratory and field. Ecology 74:1813–25.

Crowder, M. J. 1978. Beta-binomial ANOVA for proportions. Applied Statistics 27:34–37.

DeGroot, M. 1970. *Optimal Statistical Decisions.* McGraw Hill, New York.

deLeeuw, J. 1992. Introduction to Akaike (1973) information theory and an extension of the maximum likelihood principle. In *Breakthroughs in Statistics*, S. Kotz and N. Johnson (editors), 1:599–609. Springer Verlag, New York.

Deriso, R. B. 1980. Harvesting strategies and parameter estimation for an age-structured model. Canadian Journal of Fisheries and Aquatic Sciences 37:268–82.

Draper, N. R., and H. Smith. 1981. *Applied Regression Analysis.* 2nd ed. John Wiley and Sons, New York.

Dutilleul, P. 1993. Spatial heterogeneity and the design of ecological field experiments. Ecology 74:1646–58.

Edwards, A. W. F. 1992. *Likelihood.* Johns Hopkins University Press, Baltimore, Md.

Efron, B., and R. Tibshirani. 1991. Statistical data analysis in the computer age. Science 253:390–95.

Efron, B., and R. Tibshirani. 1993. *An Introduction to the Bootstrap.* Chapman and Hall, New York.

Emlen, J. M. 1989. Terrestrial population models for ecological risk assessment: a state of the art review. Environmental Toxicology and Chemistry 8:831–42.

Fagerström, T. 1987. On theory, data and mathematics in ecology. Oikos 50:258–61.

Feldman, R. M., G. L. Curry, and T. E. Wehrly. 1984. Statistical procedure for validating a simple population model. Environmental Entomology 13:1446–51.

Feller, W. 1968. *An Introduction to Probability Theory and Its Applications.* Vol. I. Wiley Interscience, New York.

Feller, W. 1971. *An Introduction to Probability Theory and Its Applications.* Vol. II. Wiley Interscience, New York.

Fernandez-Duque, E., and C. Valeggia. 1994. Meta-analysis: a valuable tool in conservation research. Conservation Biology 8:555–61.

Feynman, R. P. 1965. *The Character of Physical Law.* MIT Press, Massachusetts Institute of Technology, Cambridge, Mass.

Feynman, R. P. 1985. *Surely You're Joking Mr. Feynman.* W. W. Norton and Sons, New York.

Finn, L. S. 1994. Observational constraints on the neutron star mass distribution. Physical Review Letters 73:1878–81.

Gauch, H. G. 1993. Prediction, parsimony and noise. American Scientist. 81:468–78.

Gelman, A., J. B. Carlin, H. S. Stern, and D. B. Rubin. 1995. *Bayesian Data Analysis.* Chapman and Hall, New York.

Ghosh, J. K. 1988. *Statistical Information and Likelihood. A Collection of Critical Essays by Dr. D. Basu.* Lecture Notes in

Statistics, vol. 45. Springer Verlag, Heidelberg and New York.

Godfray, H. C. J., L. Partridge, and P. H. Harvey. 1991. Clutch size. Annual Review of Ecology and Systematics 22: 409–29.

Good, I. J. 1995. When batterer turns murderer. Nature 375:541.

Greenwood, J. J. D. 1993. Statistical power. Animal Behavior 46:1011.

Grzimek, B., and M. Grzimek. 1960. *Serengeti Shall Not Die.* Hamish Hamilton, London.

Hairston, N. G. 1989. *Ecological Experiments.* Cambridge University Press, New York.

Hairston, N. G. 1994. *Vertebrate Zoology. An Experimental Field Approach.* Cambridge University Press, New York.

Hall, C. A. S. 1988. An assessment of several of the historically most influential theoretical models used in ecology and of the data provided in their support. Ecological Modelling 43:5–31.

Hilborn, R., and A. R. E. Sinclair. 1979. A simulation of the wildebeest population, other ungulates and their predators. In Sinclair and Norton-Griffiths (1979), pp. 287–309.

Hilborn, R., and C. J. Walters. 1992. *Quantitative Fisheries Stock Assessment: Choice, Dynamics and Uncertainty.* Chapman and Hall, New York.

Hiramatsu, K., and S. Kitada. 1991. Model selection of single release tagging studies: the effect of natural mortality. Nippon Suisan Gakkaishi 57:977.

Hiyama, Y., and T. Kitahara. 1993. Relationship between surplus energy and body weight in fish populations. Researches in Population Ecology 35:139–50.

Hogarth, R. M. 1980. *Judgement and Choice. The Psychology of Decision.* John Wiley and Sons, Chichester and New York.

Hongzhi, A. 1989. Fast stepwise procedures of selection of variables by using AIC and BIC criteria. Acta Mathematicae Applicatae Sinica 5:60–67.

Hosmer, D. W., and S. Lemeshow. 1989. *Applied Logistic Regression*. Wiley Interscience, New York.

Howson, C., and P. Urbach. 1989. *Scientific Reasoning: The Bayesian Approach*. Open Court Press, La Salle, Ill.

Howson, C., and P. Urbach. 1991. Bayesian reasoning in science. Nature 350:371–74.

Huber, P. J. 1981. *Robust Statistics*. Wiley, New York.

Hudson, D. J. 1971. Interval estimation from the likelihood function. Journal of the Royal Statistical Society 33:256–62.

Hutchings, J., and R. Myers. 1994. What can be learned from the collapse of a renewable resource? Atlantic cod, *Gadus morhua*, of Newfoundland and Labrador. Canadian Journal of Fisheries and Aquatic Sciences 51:2126–46.

Jassby, A. D., and T. M. Powell. 1990. Detecting changes in ecological time series. Ecology 71:2044–52.

Jeffreys, H. 1948. *The Theory of Probability*. Clarendon Press, Oxford, U.K.

Kendall, M., and A. Stuart. 1979. *The Advanced Theory of Statistics, Volume 2. Inference*. Charles Griffin and Company, London.

Koopman, B. O. 1980. *Search and Screening*. Pergamon Press, New York.

Kramer, K. 1994. Selecting a model to predict the onset of growth of *Fagus sylvatica*. Journal of Applied Ecology 31:172–81.

Krebs, C. J. 1994. *Ecology*. 4th ed. Harper and Collins, New York.

Kuhn, T. 1962. *The Structure of Scientific Revolutions*. University of Chicago Press, Chicago, Ill.

Lack, D. 1946. Clutch and brood size in the robin. British Birds 39:98–109, 130–35.

Lack, D. 1947. The significance of clutch size. Ibis 89:302–52, 90:25–45.

Lack, D. 1948. Further notes on clutch and brood size in the robin. British Birds 41:98–137.

REFERENCES

Lakatos, I. 1978. *The Methodology of Scientific Research Programmes.* Cambridge University Press, New York.

Legendre, P. 1993. Spatial autocorrelation: trouble or new paradigm? Ecology 74:1659–73.

Leggett, R. W., and L. R. Williams. 1981. A reliability index for models. Ecological Modelling 13:303–12.

Levin, S. A., T. G. Hallam, and L. J. Gross. 1989. *Applied Mathematical Ecology.* Springer Verlag, New York.

Levins, R. 1966. The strategy of model building in population biology. American Scientist 54:421–31.

Lindh, A. G. 1993. Did Popper solve Hume's problem? Nature 366:105–6.

Linhart, H., and W. Zucchini. 1986. *Model Selection.* John Wiley and Sons, New York.

Loehle, C. 1983. Evaluation of theories and calculation tools in ecology. Ecological Modelling 19:239–47.

Ludwig, D., and C. J. Walters. 1985. Are age structured models appropriate for catch-effort data? Canadian Journal of Fisheries and Aquatic Sciences 42:1066–72.

Maddox, J. 1994. Star masses and Bayesian probability. Nature 371:649.

Mallows, C. L. 1973. Some comments on $C_p$. Technometrics 15:661–75.

Mangel, M. 1982. Applied mathematicians and naval operators. SIAM Review 24:289–300.

Mangel, M. 1987. Oviposition site selection and clutch size in insects. Journal of Mathematical Biology 25:1–22.

Mangel, M. 1992. Comparative analyses of the effects of high seas driftnets on the Northern Right Whale Dolphin *Lissodelphus Borealis.* Ecological Applications 3:221–29.

Mangel, M., and J. H. Beder. 1985. Search and stock depletion: theory and applications. Canadian Journal of Fisheries and Aquatic Sciences 42:150–63.

Mangel, M., and C. W. Clark. 1988. *Dynamic Modeling in Behavioral Ecology.* Princeton University Press, Princeton, N.J.

301

Mangel, M., and D. Ludwig. 1992. Definition and evaluation of behavioral and developmental programs. Annual Review of Ecology and Systematics 23:507–36.

Mangel, M., and F. R. Adler. 1994. Construction of multidimensional clustered patterns. Ecology 75:1289–98.

Mangel, M., J. A. Rosenheim, and F. Adler. 1994. Clutch size, offspring performance, and intergenerational fitness. Behavioral Ecology 5:412–17.

Mankin, J. B., R. V. O'Neill, H. H. Shugart, and B. W. Rust. 1975. The importance of validation in ecosystem analysis. In *New Directions in the Analysis of Ecological Systems, Part 1* (George S. Innis editor), Simulation Councils Proceedings Series 5:63–71. Simulation Councils, Inc., La Jolla, Calif.

Marten, G. G., P. M. Kleiber, and J. A. K. Reid. 1975. A computer program for fitting tracer kinetic and other differential equations to data. Ecology 56:752–54.

Martz, H., and R. Waller. 1982. *Bayesian Reliability Analysis.* John Wiley and Sons, New York.

Matsumiya, Y. 1990. AIC Introduced into Schnute's models by the removal method. Nippon Suisan Gakkaishi 56:543.

McAllister, M. K., E. K. Pikitch, A. E. Punt, and R. Hilborn. 1994. A Bayesian approach to stock assessment and harvest decisions using the sampling/importance resampling algorithm. Canadian Journal of Fisheries and Aquatic Sciences 51:2673–87.

McCullagh, P., and J. A. Nelder. 1989. *Generalized Linear Models.* Chapman and Hall, New York.

Melzer, D. A. 1970. The regression of Log $N_{n+1}$ on Log $N_n$ as a test of density dependence: an exercise with computer-constructed density-independent populations. Ecology 51:810–22.

Mills, L. S., M. E. Soule, and D. F. Doak. 1993. The keystone-species concept in ecology and conservation. BioScience 43:219–24.

Mitchell, W. A., and T. J. Valone. 1990. The optimization research program: studying adaptations by their function. Quarterly Review of Biology 65:43–52.

Morse, P. M. 1977. *In at the Beginnings: A Physicist's Life.* MIT Press, Cambridge, Mass.

Naylor, T. H., and J. M. Finger. 1967. Verification of computer simulation models. Management Science 14B:92–101.

Neufeldt, V., and D. B. Guralnik (editors). 1991. *Webster's New World Dictionary. Third College Edition.* Simon and Schuster, New York.

Nishi, R. 1984. Asymptotic properties of criteria for selection of variables in multiple regression. The Annals of Statistics 12:758–65.

Onstad, D. W. 1988. Population-dynamic theory: the roles of analytical, simulation and supercomputer models. Ecological Modelling 43:111–24.

Oreskes, N., K. Schrader-Frechette, and K. Belitz. 1994. Verification, validation and confirmation of numerical models in the earth sciences. Science 263:641–46.

Orzack, S. H. 1993. Sex ratio evolution in parasitic wasps. In *Evolution and Diversity of Sex Ratio*, D. L. Wrensch and M. A. Ebbert (editors), pp. 477–503. Chapman and Hall, New York.

Orzack, S. H., and E. Sober. 1994. Optimality models and the test of adaptationism. American Naturalist 143:361–80.

Peterman, R. M. 1990a. Statistical power analysis can improve fisheries research and management. Canadian Journal of Fisheries and Aquatic Sciences 47:2–15.

Peterman, R. M. 1990b. The importance of reporting statistical power: the forest decline and acidic deposition example. Ecology 71:2024–27.

Peters, R. H. 1991. *A Critique for Ecology.* Cambridge University Press, Cambridge, U.K.

Platt, J. R. 1964. Strong Inference. Science 146:347–53.

Polanyi, M. 1969. The republic of science: its political and economic theory. In *Knowing and Being*, Marjorie Greene (editor), pp. 50–72. University of Chicago Press, Chicago, Ill.

Popper, K. 1979. *Objective Knowledge*. Cambridge University Press, New York.

Potvin, C., and D. A. Roff. 1993. Distribution-free and robust statistical methods: viable alternatives to parametric statistics? Ecology 74:1617–28.

Potvin, C., and J. Travis. 1993. Concluding remarks: a drop in the ocean. Ecology 74:1674–76.

Press, W. H., B. P. Flannery, S. A. Teukolsky, and W. T. Vetterling. 1986. *Numerical Recipes*. Cambridge University Press, Cambridge, U.K.

Pulliam, R., and N. M. Haddad. 1994. Human population growth and the carrying capacity concept. Bulletin of the Ecological Society of America 75:141–57.

Punt, A. E. 1988. Model selection for the dynamics of southern African hake resources. M.Sc. thesis, Department of Applied Mathematics, University of Capetown. 395 pp.

Reader, R. J., S. D. Wilson, J. W. Belcher, I. Wisheu, P. A. Keddy, D. Tilman, E. C. Morris, J. B. Grace, J. B. McGraw, H. Olff, R. Turkington, E. Klein, Y. Leung, B. Shipley, R. van Hulst, M. E. Johansson, C. Nilsson, J. Gurevitch, K. Grigulis, and B. E. Beisner. 1994. Plant competition in relation to neighbor biomass: an intercontinental study with *Poa pratensis*. Ecology 75:1753–60.

Reckhow, K. H. 1990. Bayesian inference in non-replicated ecological studies. Ecology 71:2053–59.

Ribbens, E., J. A. Silander, and S. W. Pacala. 1994. Seedling recruitment in forests: calibrating models to predict patterns of tree seedling dispersion. Ecology 75:1794–806.

Ripley, B. D. 1987. *Stochastic Simulation*. Wiley Interscience, New York.

Roitberg, B. D., M. Mangel, R. Lalonde, C. A. Roitberg, J. J. M. van Alphen, and L. Vet. 1992. Seasonal dynamic shifts in patch exploitation by parasitic wasps. Behavioural Ecology 3:156–65.

Roitberg, B. D., J. Sircom, C. A. Roitberg, J. J. M. van Alphen, and M. Mangel. 1993. Life expectancy and reproduction. Nature 364:351.

Rosenheim, J. A., and D. Rosen. 1991. Foraging and oviposition decisions in the parasitoid *Aphytis lingnanensis:* distinguishing the influences of egg load and experience. Journal of Animal Ecology 60:873–93.

Rosenheim, J. A., and M. Mangel. 1994. Patch-leaving rules for parasitoids with imperfect host discrimination. Ecological Entomology 19:374–80.

Roughgarden, J., T. Pennington, and S. Alexander. 1994. Dynamics of the rocky intertidal zone with remarks on generalization in ecology. Philosophical Transactions of the Royal Society of London B 343:79–85.

Rousseeuw, P. J. 1984. Least median of squares regression. Journal of the American Statistical Association 79:871–80.

Sakamoto, Y., M. Ishiguro, and G. Kitagawa. 1986. *Akaike Information Statistics.* KTK Scientific Publishers, Tokyo, and D. Reidel Publishing, Dordrecht.

Samaniego, F. J., and D. M. Reneau. 1994. Toward a reconciliation of the Bayesian and frequentist approaches to point estimation. Journal of the American Statistical Association 89:947–57.

Sanderson, M. J. and M. J. Donoghue. 1994. Shifts in the diversification rate with the origin of angiosperms. Science 264:1590–93.

Santer, B. D., and T. M. L. Wigley. 1990. Regional validation of means, variances, and spatial runs in general circulation model control runs. Journal of Geophysical Research 95D1:829–950.

Schnute, J. T. 1987. Data, uncertainty, model ambiguity, and model identification. Natural Resource Modeling 2:159–212.

Schnute, J. T. 1993. Ambiguous inferences from fisheries data. In *Statistics for the Environment,* V. Barnett and K. F.

Turkman (editors), pp. 293–309. John Wiley and Sons, Ltd., London and New York.

Schnute, J. T., and K. Groot. 1992. Statistical analysis of animal orientation data. Animal Behavior 43:15–33.

Schnute, J. T., and R. Hilborn. 1993. Analysis of contradictory data sources in fish stock assessment. Canadian Journal of Fisheries and Aquatic Sciences 50:1916–23.

Schwarz, G. 1978. Estimating the dimension of a model. Annals of Statistics 2:461–64.

Seber, G. A. F. 1980. *The Estimation of Animal Abundance.* Macmillan Press, New York.

Selzer, J. 1993. *Understanding Scientific Prose. Rhetoric of the Human Sciences.* University of Wisconsin Press, Madison, Wis.

Shaeffer, D. L. 1980. A model evaluation methodology applicable to environmental assessment models. Ecological Modelling 8:275–95.

Shaver, J. P. 1993. What statistical significance testing is and what it is not. Journal of Experimental Education 61:293–316.

Shaw, R. G., and T. Mitchell-Olds. 1993. ANOVA for unbalanced data: An overview. Ecology 74:1638–45.

Sheail, J. 1989. Obituary. Professor Sir Harold Jeffreys. Bulletin of the British Ecological Society 20:200–202.

Shrader-Frechette, K. S., and E. D. McCoy. 1992. Statistics, costs and rationality in ecological inference. Trends in Ecology and Evolution 7:96–99.

Sinclair, A. R. E. 1979. The eruption of ruminants. In Sinclair and Norton-Griffiths (1979), pp. 82–103.

Sinclair, A. R. E., and M. Norton-Griffiths. 1979. *Serengeti: Dynamics of an Ecosystem.* University of Chicago Press, Chicago, Ill.

Sinclair, A. R. E., and P. Arcese. 1995. *Serengeti II. Research, Management and Conservation of an Ecosystem.* University of Chicago Press, Chicago, Ill.

Smith, T. D. 1994. *Scaling Fisheries.* Cambridge University Press, New York.

Sokal, R. R., and F. J. Rohlf. 1969. *Biometry*. W.H. Freeman and Company, San Francisco.

Southwood, T. R. E. 1966. *Ecological Methods*. Chapman and Hall, London.

Speed, T. 1993. Modelling and managing a salmon population. In *Statistics for the Environment*, V. Barnett and K. F. Turkman (editors), pp. 267–92. John Wiley and Sons, Ltd., London and New York.

Stephens, D., and J. R. Krebs. 1986. *Foraging Theory*. Princeton University Press, Princeton, N.J.

Stigler, S. M. 1986. *The History of Statistics. The Measurement of Uncertainty Before 1900*. Harvard University Press, Cambridge, Mass.

Stow, C. A., S. R. Carpenter, and K. L. Cottingham. 1995. Resource vs. ratio-dependent consumer-resource models: A Bayesian perspective. Ecology 76:1986–90.

Tanzania Wildlife Conservation Monitoring. 1994. *Status and Trends of Wildebeest in the Serengeti Ecosystem*. Frankfurt Zoological Society, POB 3134, Arusha, Tanzania.

Thompson, C. F., and A. J. Neill. 1993. Statistical power and accepting the null hypothesis. Animal Behavior 46:1012.

Thompson, G. G. 1992. A Bayesian approach to management advice when stock-recruitment parameters are uncertain. Fishery Bulletin, U.S. 90:561–73.

Tidman, K. R. 1984. *The Operations Evaluation Group*. Naval Institute Press, Annapolis, Md. 388 pp.

Trexler, J. C., and J. Travis. 1993. Nontraditional regression analyses. Ecology 74:1629–37.

Tuckwell, H., and J. Koziol. 1992. World population. Nature 359:200.

Ulanowicz, R. E. 1988. On the importance of higher-level models in ecology. Ecological Modelling 43:45–56.

Underwood, A. J. 1991. Experiments in ecology and management: their logics, functions and interpretations. Australian Journal of Ecology 15:365–89.

Venzon, D. J., and S. H. Moolgavkar. 1988. A method for

computing profile-likelihood based confidence intervals. Applied Statistics 37:87–94.

Walters, C. J. 1986. *Adaptive Management of Renewable Resources.* Macmillan Publishing Company, New York.

Walters, C. J., and C. S. Holling. 1990. Large-scale management experiments and learning by doing. Ecology 71: 2060–68.

Walters, C. J., and A. E. Punt. 1994. Placing odds on sustainable catch using Virtual Population Analysis and survey data. Canadian Journal of Fisheries and Aquatic Sciences 51:946–58.

Wigley, T. M. L., and B. D. Santer. 1990. Statistical comparison of spatial fields in model validation, perturbation, and predictability experiments. Journal of Geophysical Research 95D1:851–65.

Wise, D. H. 1993. *Spiders in Ecological Webs.* Cambridge University Press, Cambridge, U.K.

Wismer, D., and R. Chattergy. 1978. *Introduction to Nonlinear Optimization.* North-Holland, New York.

# Index

age-structured models, 242
aggregated data, 98–104
AIC. *See* Akaike information criterion
Akaike information criterion (AIC), 159–60; fisheries example of, 255–56, 260–62
analysis of deviance, 160
analysis of variance: by likelihood methods, 137, 171–72, 177–79
ANOVA. *See* analysis of variance
assumptions: of model, 30; about sources of uncertainty, 131, 150–51

BASIC, 277
Bayesian analysis, 8–9; advantages of, 203–4; vs. classical statistics, 19–21; controversies about, 9, 205–6; of discrete examples, 206–12; of examples with binomial distribution, 221–33; of example with gamma distribution, 220–21; of example with Poisson distribution, 214–20; of fisheries models, 256–60, 261–62; vs. likelihood methods, 212, 231–33; with limited data, 212, 228; with limited prior information, 212–14, 231–33; model selection with, 117, 233–34, 261–62. *See also* prior probabilities
Bayesian confidence intervals, 223–24, 226, 230–31
Bayesian information criterion (BIC), 116, 160
Bayes' theorem, 43–47, 204–5
best-fit parameters, 148
beta-binomial model, 223
beta density, 223
Beverton-Holt stock recruitment curve, 245–46

bias, observation, 60–61, 152
BIC (Bayesian information criterion), 116, 160
binomial distribution, 62–63, 64–67; Bayesian analysis with, 221–33; generating random variables with, 88–89
biomass dynamics models, 242
birds. *See* incidental catch
biweight: for outliers, 161–62
bootstrap method, 92–93; advantage and disadvantage of, 130; computation time for, 171; for finding confidence intervals and variances, 168–71; model selection with, 116–17, 128–29; with random variable added, 171
Box-Mueller scheme, 89
by-catch. *See* incidental catch

calculus facts, 51–52
carrying capacity in logistic model, 146; difficulty of estimating, 193–94, 202
catch per unit effort (CPUE), 237–41, 248–50, 254, 257, 261
central limit theorem, 63, 73
chain rule, 52
Chamberlain, T. C., 13, 281
chaotic models, 151–52
chi-square distribution, 76
chi-square test: of maximum likelihood estimate, 154–55, 173–74; of negative binomial model, 101, 103
classical hypothesis testing, 6–7, 9; vs. Bayesian analysis, 19–21; statistical theory of, 14–15
clumped data, 58, 70
clutch size. *See* oviposition behavior

309